ASCE Manuals and Reports on Engineering Practice No. 96

Guide to Improved Earthquake Performance of Electric Power Systems

Prepared by
Electric Power and Communications Committee
Technical Council on Lifeline Earthquake Engineering

Edited by
Anshel J. Schiff

Published by
ASCE American Society of Civil Engineers
1801 Alexander Bell Drive
Reston, Virginia 20191-4400

Abstract: The Manual documents methods to improve the earthquake response of electric power systems. A review of the Manual should raise the awareness and understanding of the vulnerabilities of power system facilities and equipment. The emphasis is on power system elements that have been damaged by earthquakes, primarily high-voltage substation equipment. Power-generating stations, transmission and distribution lines, substations, system communications and control, and ancillary facilities and functions are also discussed. A detailed review of earthquake damage to power facilities and suggestions to improve their performance are presented. The Manual suggests an overall approach to an earthquake mitigation program. Postearthquake emergency response procedures to reduce the disruption from damaged facilities are also discussed. The Manual demonstrates that improved installation practices and other mitigation measures, particularly for new construction and during refurbishment, that are cost-effective in any region with a history of significant earthquakes can be implemented to improve earthquake performance.

Library of Congress Cataloging-in-Publication Data

American Society of Civil Engineers. Electric Power and Communications Committee.
Guide to improved earthquake performance of electric power systems / prepared by Electric Power and Communications Committee, Technical Council on Lifeline Earthquake Engineering.
p. cm.—(ASCE manuals and reports on engineering practice ; no. 96)
Includes bibliographical references and index.
ISBN 0-7844-0414-3
1. Electric power systems—Earthquake effects. 2. Lifeline earthquake engineering. I. Title. II . Series.
TK1005.A633 1999
621.31′21—dc21 99-20666
CIP

Library of Congress Catalog Card No: 99-20666
ISBN 0-7844-0414-3
Manufactured in the United States of America

MANUALS AND REPORTS ON ENGINEERING PRACTICE

(As developed by the ASCE Technical Procedures Committee, July 1930, and revised March 1935, February 1962, and April 1982)

A manual or report in this series consists of an orderly presentation of facts on a particular subject, supplemented by an analysis of limitations and applications of these facts. It contains information useful to the average engineer in his everyday work, rather than the findings that may be useful only occasionally or rarely. It is not in any sense a "standard," however; nor is it so elementary or so conclusive as to provide a "rule of thumb" for non-engineers.

Furthermore, material in this series, in distinction from a paper (which expressed only one person's observations or opinions), is the work of a committee or group selected to assemble and express informaton on a specific topic. As often as practicable the committee is under the direction of one or more of the Technical Divisions and Councils, and the product evolved has been subjected to review by the Executive Committee of the Division or Council. As a step in the process of this review, proposed manuscripts are often brought before the members of the Technical Divisions and Councils for comment, which may serve as the basis for improvement. When published, each work shows the names of the committees by which it was compiled and indicates clearly the several processes through which it has passed in review, in order that its merit may be definitely understood.

In February 1962 (and revised in April 1982) the Board of Direction voted to establish:

> A series entitled "Manuals and Reports on Engineering Practice," to include the Manuals published and authorized to date, future Manuals of Professional Practice, and Reports on Engineering Practice. All such Manual or Report material of the Society would have been refereed in a manner approved by the Board Committee on Publications and would be bound, with applicable discussion, in books similar to past Manuals. Numbering would be consecutive and would be a continuation of present Manual numbers. In some cases of reports of joint committees, bypassing of Journal publications may be authorized.

MANUALS AND REPORTS OF ENGINEERING PRACTICE

No.	*Title*
13	Filtering Materials for Sewage Treatment Plants
14	Accommodation of Utility Plant Within the Rights-of-Way of Urban Streets and Highways
34	Definitions of Surveying and Associated Terms
35	A List of Translations of Foreign Literature on Hydraulics
37	Design and Construction of Sanitary and Storm Sewers
40	Ground Water Management
41	Plastic Design in Steel: A Guide and Commentary
45	Consulting Engineering: A Guide for the Engagement of Engineering Services
46	Pipeline Route Selection for Rural and Cross-Country Pipelines
47	Selected Abstracts on Structural Applications of Plastics
49	Urban Planning Guide
50	Planning and Design Guidelines for Small Craft Harbors
51	Survey of Current Structural Research
52	Guide for the Design of Steel Transmission Towers
53	Criteria for Maintenance of Multilane Highways
54	Sedimentation Engineering
55	Guide to Employment Conditions for Civil Engineers
57	Management, Operation and Maintenance of Irrigation and Drainage Systems
59	Computer Pricing Practices
60	Gravity Sanitary Sewer Design and Construction
62	Existing Sewer Evaluation and Rehabilitation
63	Structural Plastics Design Manual
64	Manual on Engineering Surveying
65	Construction Cost Control
66	Structural Plastics Selection Manual
67	Wind Tunnel Studies of Buildings and Structures
68	Aeration: A Wastewater Treatment Process
69	Sulfide in Wastewater Collection and Treatment Systems
70	Evapotranspiration and Irrigation Water Requirements
71	Agricultural Salinity Assessment and Management
72	Design of Steel Transmission Pole Structures
73	Quality in the Constructed Project: A Guide for Owners, Designers, and Constructors
74	Guidelines for Electrical Transmission Line Structural Loading
75	Right-of-Way Surveying
76	Design of Municipal Wastewater Treatment Plants
77	Design and Construction of Urban Stormwater Management Systems
78	Structural Fire Protection
79	Steel Penstocks
80	Ship Channel Design
81	Guidelines for Cloud Seeding to Augment Precipitation
82	Odor Control in Wastewater Treatment Plants
83	Environmental Site Investigation
84	Mechanical Connections in Wood Structures
85	Quality of Ground Water
86	Operation and Maintenance of Ground Water Facilities
87	Urban Runoff Quality Manual
88	Management of Water Treatment Plant Residuals
89	Pipeline Crossings
90	Guide to Structural Optimization
91	Design of Guyed Electrical Transmission Structures
92	Manhole Inspection and Rehabilitation
93	Crane Safety on Construction Sites
94	Inland Navigation: Locks, Dams, and Channels
95	Urban Subsurface Drainage
96	Guide to Improved Earthquake Performance of Electric Power Systems
97	Hydraulic Modeling: Concepts and Practice

TABLE OF CONTENTS

PREFACE

The development of this document began with a grant from the National Science Foundation with subsequent support from the Electric Power Research Institute and the National Institute of Standards and Technology. The technical content of the document has evolved as additional information has been gathered following damaging earthquakes. The lessons learned from the Northridge earthquake were particularly useful.

Unlike the more frequent disasters that impact power systems, such as severe winds and ice storms, damaging earthquakes may not impact a utility in less seismically active areas during the entire working careers of its personnel. Further, many utilities in seismically vulnerable areas have not been subjected to a damaging earthquake since the introduction of modern power systems. Thus, there is little opportunity for a meaningful learning curve to develop within an organization. Executive Order 12699 mandates that all federal organizations or organizations receiving federal funds must consider seismic vulnerability in the design and construction of their buildings. Under Executive Order 12941, such organizations must review the seismic vulnerability of existing buildings. This guide has been written to address the need for practical guidance for improving earthquake performance of power facilities.

A special effort has been made to have broad representation in the review of this document. Active participants in the review process included representatives from the most seismically aggressive utilities; investor-owned, municipal, and federal power organizations; West Coast and Eastern utilities; and engineer-architects and consultants who design power facilities. Members of the committee and organizations not represented on the committee were asked and participated in the review of the document. All individuals were asked to accept, accept with reservations, or reject the document on a mailed ballot, and all accepted without reservations.

Anshel J. Schiff

ACKNOWLEDGMENTS

The cooperation of the Los Angeles Department of Water and Power (LADWP), the Pacific Gas and Electric Company (PG&E), and Southern California Edison (SCE) for facilitating postearthquake investigation of their facilities is gratefully acknowledged. Without the information gained from the California and foreign earthquakes, the preparation of this document would not have been possible.

The support of the National Institute of Standards and Technology, the Electric Power Research Institute, and the National Science Foundation for the preparation of this document is also acknowledged. We are grateful to the American Society of Civil Engineers (ASCE), the National Center for Earthquake Engineering Research, the Federal Emergency Management Agency, and the National Academy of Sciences for their contributions for travel expenses for postearthquake investigations. Without this support, investigations could not have been as thorough, and many international investigations would not have been possible.

The author gratefully acknowledges the many technical discussions and the generous and thoughtful sharing of expertise of Rulon Fronk (LADWP), Dennis Ostrom (formerly SCE), Edward Matsuda (PG&E), and Ron Tognazzini (LADWP) in the course of postearthquake investigations.

The support of the Earthquake Investigation Committee of the Technical Council on Lifeline Earthquake Engineering, ASCE, and its postearthquake investigators, who have gathered and interpreted earthquake damage data over the years, has contributed to this document. The volunteer activities of these and other investigators were primarily self-supported.

The many technical contributions and constructive suggestions of the reviews are acknowledged. The primary contributors are listed below.

Eric Fujisaki, Pacific Gas and Electric
Rulon Fronk, Los Angeles Department of Water and Power

Joseph Graziano, Tennessee Valley Authority
Leon Kempner, Jr., Bonneville Power Administration
Jim Kennedy, Southern California Edison
Edward Matsuda, Pacific Gas and Electric
Robert McBean, Black and Veatch
Dennis Ostrom, consultant and formerly with Southern California Edison
Otto Steinhardt, formerly with Pacific Gas and Electric
Alex Tang, Nortel
Ron Tognazzini, Los Angeles Department of Water and Power
Masoud Moghtaderi-Zadeh, K2 Technologies, Inc.

The editorial contributions of Otto Steinhardt and Susan Welch significantly improved readability of the document.

EXECUTIVE SUMMARY

Recent moderate and strong California earthquakes have demonstrated that parts of electric power systems are very vulnerable to damage. Most damage has been due to the failure of porcelain elements in high-voltage substation equipment. However, performance is strongly influenced by specific equipment designs and installation practices. There has also been damage to substation buildings, conductor support structures, cast-aluminum hardware used on both low- and high-voltage equipment, equipment support structures, equipment anchorage, and parts of power-generating stations. The performance of some communication and control systems has been impaired after earthquakes.

Direct cost for repair and replacement of earthquake damage to power system facilities from the 1971 San Fernando (Moment Magnitude, M_w = 6.6), the 1987 North Palm Springs (M_w = 6.1), the 1989 Loma Prieta (M_w = 6.9), and the 1994 Northridge (M_w = 6.7) earthquakes was \$45 million, \$9 million, \$100 million, \$183 million, respectively. In spite of extensive equipment damage, system performance after these earthquakes has been good in terms of customer disruption because of the high degree of redundancy incorporated into power systems. However, no major or great earthquake has struck a large metropolitan area. Past earthquake damage to equipment suggests that larger earthquakes, earthquakes that impact larger areas, or earthquakes that occur in regions where less stringent seismic design practices and more vulnerable equipment are used will have more extensive earthquake damage that could overwhelm system redundancies. As a result, unacceptably large direct losses, lengthy disruption of service to the community, and indirect losses borne by customers are likely. Improved installation practices and other mitigation measures, particularly for new construction and during refurbishment, that are cost-effective in any region with a history of significant earthquakes can be implemented to improve earthquake performance.

The purpose of this Manual is to document methods to improve the earthquake response of electric power systems. The major goals include:

- reviewing how earthquakes affect power system facilities and equipment;
- raising the awareness and understanding of the vulnerabilities of power system facilities and equipment by reviewing their earthquake performance;
- suggesting an overall approach to an earthquake mitigation program;
- suggesting earthquake preparedness techniques to improve postearthquake response;
- reviewing design details that contributed to both good performance and failures during earthquakes;
- providing insight into facility performance, so facilities can be evaluated to determine their earthquake vulnerability;
- suggesting hardware changes that can reduce earthquake damage to existing facilities;
- suggesting approaches for new construction that have been shown to reduce earthquake damage; and
- suggesting earthquake emergency response procedures to reduce the disruption from damaged facilities.

This document deals with major power system elements—power-generating stations, transmission and distribution lines, substations, system communications and control, and ancillary facilities and functions. The emphasis given to the various elements is strongly related to their earthquake performance. Thus, a large portion of the document is devoted to high-voltage substations, as this is where most power system damage has been concentrated. In facilities where existing practices have performed well, such as transmission lines and power-generating stations, recommendations will be limited to those areas where damage has been observed.

Key findings of the document are the following.

- It is vital to start an earthquake mitigation program.
- Cost-effective methods are available to improve the earthquake performance of power systems.
- Earthquake mitigation practices should be institutionalized by incorporating them into the utility's manual of practice.
- Most seismic upgrading of a utility will occur during normal refurbishment and new construction and may require a few decades to be fully implemented.

In most earthquakes, it has been possible to bypass damaged equipment and continue to get power through or to route power around the damaged substation. There have been cases where an entire switchyard has been bypassed. Fortunately, when transformers have been damaged, there has been adequate capacity in alternate routes to maintain service. However, it is easy to envision damage, particularly to transformers, that could cause lengthy disruptions. The relatively short time to restore service in the face of extensive damage can be attributed to the high level of redundancy designed into power systems and the resourcefulness and dedication of utility maintenance personnel.

Looking at a utility's response to moderate and strong earthquakes helps to put the recovery effort into perspective. In the 1986 North Palm Springs earthquake, where a single substation was damaged, about 250 people worked 18-hour days for three days to clear damaged equipment from the site. About 180 people continued to work for about an additional 6.5 days to restore service to a critical line. This reconstruction was carried out by several crews working in parallel at all locations where possible, to reduce the disruption time. This damage occurred to a facility that used the then-current and most stringent earthquake mitigation practices instituted after the 1971 San Fernando earthquake. After the Loma Prieta and Northridge earthquakes, it took months to repair and replace damaged equipment, even though service was restored quickly.

In light of the recent experience, many California utilities are reevaluating the vulnerability of their systems and adopting measures to improve their system's response. While some of these measures would be difficult to justify in regions of lower risk, many things, particularly for new construction, can be done that are cost-effective in any region that has a history of significant earthquakes. It would be unfortunate if cost-effective measures were not implemented and that utilities and the communities they serve were exposed to avoidable risks.

Chapter 1
INTRODUCTION

1.1 BACKGROUND

Recent moderate and strong California earthquakes have demonstrated that parts of power systems are very vulnerable to damage. While system performance has been good as measured by customer disruption, the damage pattern suggests that system performance will not be acceptable for larger earthquakes, for earthquakes that impact larger areas, or for earthquakes in regions of the country that use less stringent earthquake practices. Starting with the 1971 San Fernando earthquake, there have been ten California earthquakes that have damaged power system facilities. Most of these small to moderate earthquakes have shaken relatively small areas, so that damage has been concentrated and limited primarily to one or two facilities. The 1989 Loma Prieta, California, earthquake affected a large area; three substations were severely damaged, and the network in the region was disrupted. The 1994 Northridge, California, earthquake also affected a large area, and eleven power facilities were damaged; several experienced peak ground accelerations above 0.75*g*. Earthquakes in Chile and Japan have also provided lessons that can be applied to facilities in the United States.

Most damage has been due to the failure of porcelain elements in high-voltage substation equipment. However, performance is strongly influenced by specific equipment designs and installation practices. There has also been damage to substation buildings, conductor support structures, cast aluminum hardware used on both low- and high-voltage equipment, equipment support structures, equipment anchorage, and parts of power generating stations. The performance of some communication and control systems has also been impaired after earthquakes.

Direct costs of earthquakes for repair and replacement of damaged facilities have not been overwhelming in terms of utility assets, but are nonetheless significant. There were direct losses of about $45 million in the 1971 San

Fernando earthquake (Moment Magnitude, M_W = 6.6), \$9 million in the 1986 North Palm Springs earthquake (M_W = 6.1), \$100 million in the 1989 Loma Prieta earthquake (M_W = 6.9), and \$183 million in the 1994 Northridge earthquake (M_W = 6.7). Some smaller, less costly earthquakes have been very disruptive. In the 1988 Tejon Ranch earthquake, the California Aqueduct, which supplies water to the Los Angeles area, was shut down for four days; there is typically only a 15-day supply of water stored downstream of the point that was damaged. In the 1984 Morgan Hill earthquake, one of the three Pacific Interties (major power circuits connecting the Northwest and the Southwest) was down for three days.

In most earthquakes, it has been possible to bypass damaged equipment and continue to receive power through or to route power around the damaged substation. There have been cases where an entire switchyard has been bypassed. Fortunately, when transformers have been damaged, there has been adequate capacity in alternate routes to maintain service. The relatively short time to restore service in the face of extensive damage can be attributed to the high level of redundancy designed into power systems and the resourcefulness and dedication of utility maintenance personnel. However, it is easy to envision damage, particularly to transformers, that could cause lengthy disruptions.

Looking at a utility's response to moderate earthquakes helps put the recovery effort into perspective. In the 1986 North Palm Springs earthquake, about 250 people worked 18-hour days for three days to clear damaged equipment from the site. About 180 people continued to work for about an additional 6½-days to restore equipment and service to a critical line. This reconstruction was carried out by several crews working in parallel at all locations where possible in order to reduce the disruption time. This damage occurred to a facility that used the then-current and most stringent earthquake mitigation practices. After the Loma Prieta and Northridge earthquakes, it took months to repair and replace damaged equipment, even though service was restored quickly.

This experience suggests that larger earthquakes, earthquakes that impact larger areas, or earthquakes that occur in regions where less stringent seismic design practices and more vulnerable equipment are used will have more extensive earthquake damage that could overwhelm system redundancies. As a result, unacceptably large direct losses, indirect losses borne by customers, and lengthy disruption of service to the community are likely. In light of the recent experience, large California utilities are reevaluating the vulnerability of their systems and are adopting measures to improve their system responses. While many of these measures would be difficult to justify in regions of lower risk, some things—particularly for new construction—can be done that are cost-effective in any region that has a history of significant earthquakes. It would be unfortunate if cost-effective measures were not implemented, and utilities and the communities they serve were exposed to avoidable risks.

Much of the information on earthquake damage to power systems has been gathered by the Earthquake Investigation Committee of the Technical Council on Lifeline Earthquake Engineering. Members of the committee investigated eleven U.S. and five foreign earthquakes that have damaged power system facilities, including the 1985 Chile; 1978 Sendai, Japan; 1988 Soviet Armenia; 1990 Philippines; and 1995 Kobe, Japan earthquakes.

1.2 PURPOSE

The purpose of this Manual is to improve the earthquake response of electric power systems. This is to be achieved by doing the following:

- review how earthquakes affect power system facilities and equipment;
- raise the awareness and understanding of the vulnerabilities of power system facilities and equipment by reviewing their earthquake performance;
- suggest an overall approach to an earthquake mitigation program;
- suggest earthquake preparedness techniques to improve postearthquake response;
- review design details that contributed to failures or that performed well in earthquakes;
- provide insight into facility performance so that facilities could be evaluated to determine their earthquake vulnerability;
- suggest hardware changes that could reduce earthquake damage;
- suggest approaches for new construction that have been shown to reduce earthquake damage; and
- suggest earthquake response procedures to reduce the disruption from damaged facilities.

1.3 BASIS FOR RECOMMENDATIONS

The recommendations in this Manual are based primarily on experience in California earthquakes and a few foreign earthquakes. Many of the recommendations have been drawn from innovations initiated by California utilities, and their contributions are acknowledged. Because practices and equipment used in different parts of the United States vary, the effectiveness of some practices and equipment is unknown. Also, there may be differences in earthquakes and their effects in other parts of the country. For example, there are indications from records of small, Eastern earthquakes that ground motions are richer in higher frequencies, which may result in differences in equipment performance. Also, it is known that larger areas in the Midwest and Southeast than in California are vulnerable to soil liquefaction.

Some facilities have no significant earthquake experience. For example, there is no record of a coal-fueled power-generating plant being subjected to a significant earthquake. There are no traditional coal-fired plants in California nor in the areas investigated after the 1985 Chile, 1978 Sendai, Japan, or the 1995 Kobe, Japan, earthquakes.

It is important to realize that the seismic design of power systems is a work in progress. Future earthquakes can be expected to suggest changes in practice as they have done in the past.

1.4 SCOPE

This Manual deals with major power system elements—power-generating stations, transmission and distribution lines, substations, system communications and control, and ancillary facilities and functions. The emphasis given to the various elements is strongly related to their earthquake performance. Thus, a large portion of the document is devoted to high-voltage substations, as this is where most power system damage has been concentrated. Where existing practices, even though they may not be well defined, have performed well (as, for example, in power generating plants), recommendations will be limited to those areas where damage has been observed.

This Manual is meant to identify practices that have yielded either good or poor performance and to suggest mitigation methods. Seismic design criteria are determined by the utilities. Guidance for design criteria for substation structures are contained in ASCE's *Substation Structure Design Guide* (currently being developed). The Institute of Electrical and Electronic Engineers (IEEE) Standard 693–1997, "Recommended Practices for Seismic Design of Substations," provides guidance for seismic qualification criteria and methods for substation equipment; when given time for its recommendations to find their way into the field, these recommendations should improve the seismic performance of such equipment.

1.5 ORGANIZATION OF THE MANUAL

Chapter 2 describes the sources of earthquakes and their effects with an emphasis on power system impacts. Chapter 3 reviews the overall performance of power systems and power facilities. Chapter 4 describes approaches for improving the earthquake performance of power systems. Chapters 5 through 11 deal with detailed technical issues. These chapters have been organized so that recommended practices are presented at the beginning of the technical sections. The recommendations are divided into three groups: recommendations applicable to both renovation of existing facilities and new construction, recommendations applicable primarily to the renovation of existing facilities, and recommendations applicable primarily to new construction. Material following the recommendations provides background for the recommendations. The discussion of the equipment and facilities in these chapters is typically divided into five topics. The five topics are:

1. function and brief description of equipment;
2. the earthquake performance;
3. emergency operation procedures used to mitigate damage;
4. retrofit and mitigation methods for existing facilities; and
5. methods for new construction.

Chapters 5 through 11 deal with the following topics: substations, transmission and distribution line networks, power-generating stations, system control, communication system facilities, and specialized facilities and ancillary facilities and functions. Appendix A contains information on the Modified Mercalli Intensity Scale, which is used to characterize the severity of an earthquake at a specific location. Appendix B discusses soil-structure interaction. Appendix C describes substation bus configurations. Appendix D contains resources that provide additional information about earthquake effects on power systems. Appendix E contains a list of TCLEE publications and monographs.

Chapter 2

EARTHQUAKES: SOURCES AND EFFECTS

Recommendation

Strong-motion accelographs should be installed and maintained at important facilities so that ground motion at the site can be properly characterized. (2.3.1)

2.1 SOURCES OF EARTHQUAKES

The crust of the earth, that is, the solid part near the earth's surface, is made up of large plates. These plates, for reasons not completely understood, are moving slowly relative to each other. Most earthquakes can be traced to the boundaries between these plates. In some cases, such as along the southern part of the San Andreas Fault in California, the two plates are slipping relative to each other, parallel to the fault in a horizontal plane. In other locations, such as the Aleutian Island Arc in Alaska and along the coast north of Eureka, California, one plate is slipping beneath the other (a *subduction zone*). Figure 2-1 is a map that shows earthquake locations. The plate boundaries are clearly indicated by the clustering of earthquakes. Earthquakes also occur away from the plate boundaries, because forces between the plates at the boundaries create stress fields throughout the plate. These stresses and other effects are associated with the midplate earthquakes.

A more detailed view of earthquake locations within the contiguous United States is shown in Figure 2-2. It is clear from this figure that "earthquake country" extends beyond the borders of California.

Although continual creep and slippage occur at some locations along fractures in the rock, at most locations and most of the time, the plate boundaries are locked, that is, there is no relative motion along the plate

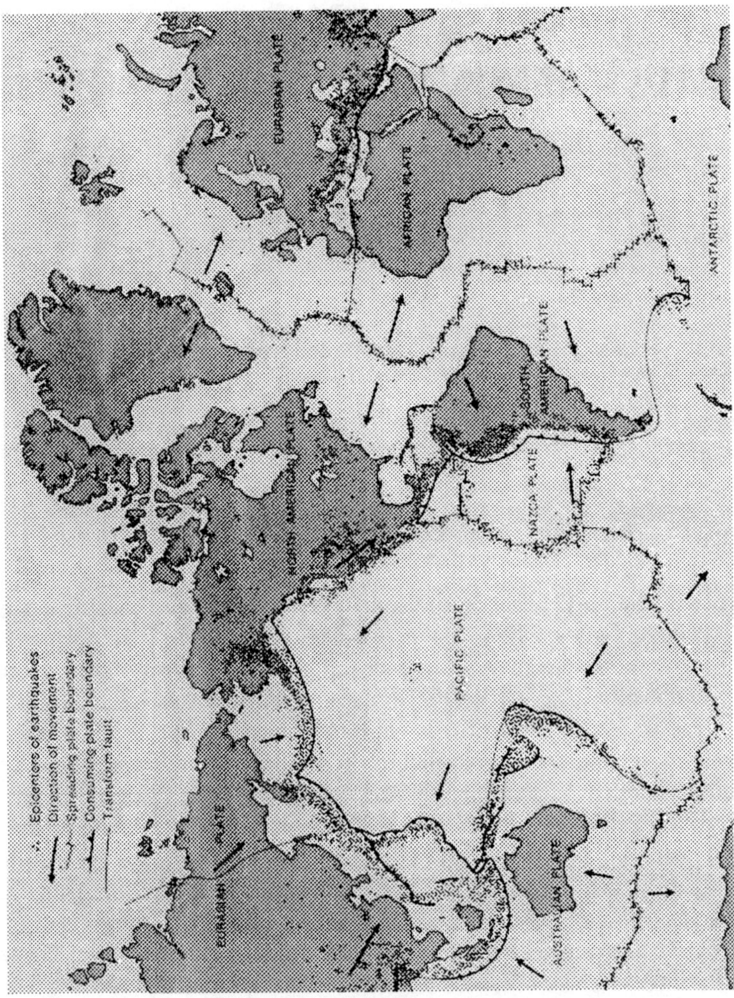

FIGURE 2-1. World Seismicity Map.

Source: Based on a U.S. Geological Survey map, "The Dynamic Planet."

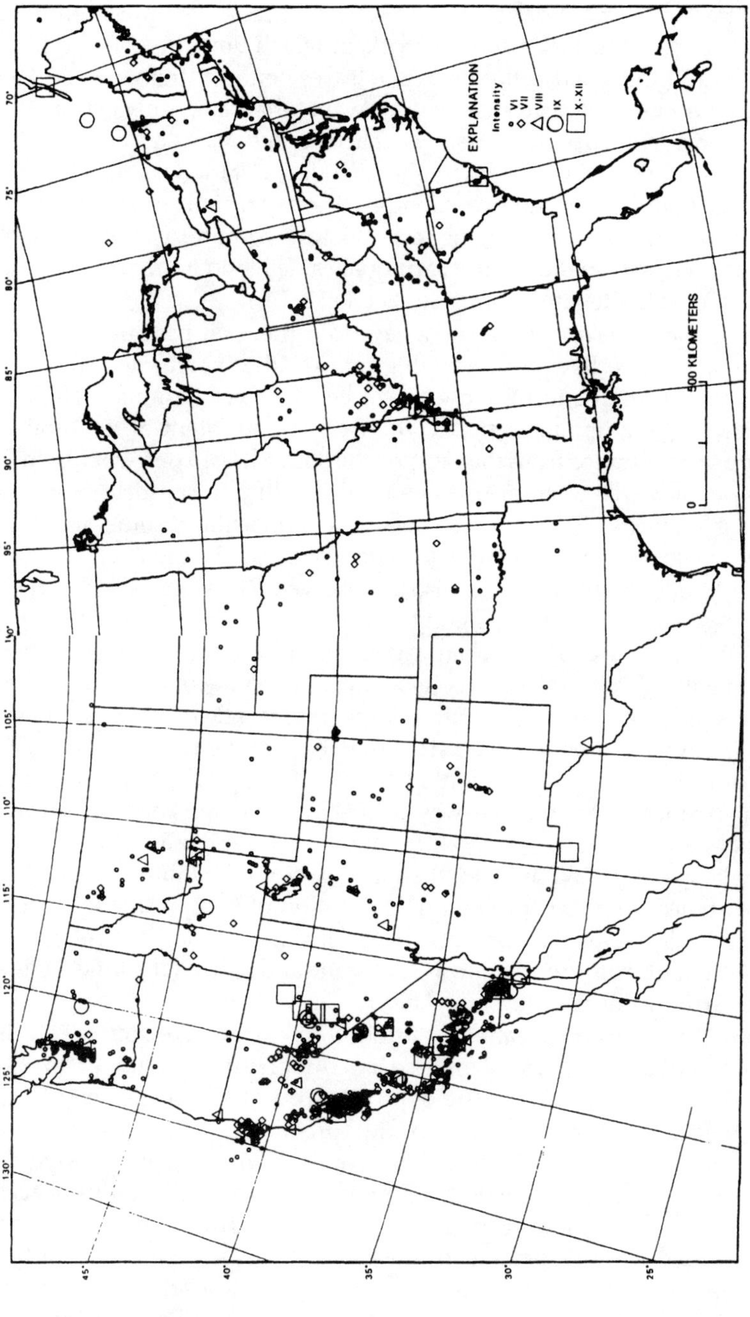

FIGURE 2-2. *Distribution of Epicenters in the United States.*

Source: C.W. Stover and J.L. Coffman. 1993. *Seismicity of the United States, 1568–1989.* USGS Professional Paper 1527.

boundaries. However, the overall motion of the plates continues, so that the stress at the locked boundaries increases. This process continues until the stress approaches the strength of the rock, at which time there is a sudden rupture and slipping across the rupture releases pent-up strain energy. The release of the strain energy creates seismic waves that radiate away from the rupture zone. The rupture surface is called a *fault*. The location on the earth's surface above the spot where the earthquake started is called the *epicenter.* If the fault fracture surface extends to the earth's surface, the fracture line is called the *fault trace*. Once the rocks fracture, forming a fault, a zone of weakness is created so that future earthquakes will tend to occur along the old fault. (For extended definitions, see Section 2.7.1.)

As the seismic waves propagate away from the fault rupture, they are attenuated. Also, as they pass through ground with different densities and physical structures, the form of the waves becomes very complex. Higher-frequency waves attenuate faster, so the frequency content of the motion changes as the distance from the fault rupture increases. There are also different types of waves. Some are *compressional,* in which the particle motion is in the same direction in which the wave is propagating; some are *shear waves,* in which the particle motion is perpendicular to the direction of propagation. Other types of waves can also be observed. The velocity with which each of these waves travels is different.

Since the stress that the rock can withstand prior to fracturing is limited, the strain energy that can be stored in any given volume of rock is also limited. Thus, a great earthquake with a large energy release requires that a large volume of rock be involved. This implies that the length of the fault involved in the event must be long. For example, the fault rupture in the 1906 San Francisco earthquake was over 400 km long. This has several important consequences. The ground motion at a site impacted by a large earthquake is influenced more by the distance from the fault rupture than by the distance from the epicenter. The duration of motion due to a great earthquake may be several minutes, since at any given location there will be parts of the fault rupture that are very distant and it will take time for the waves to arrive from all parts of the fault.

Figure 2-3 shows an idealized model of an earthquake source that is used to describe several variables associated with an earthquake.[1] In this model, the rupture begins at the hypocenter located a distance h (focal depth) below the surface. The rupture spreads across the fault plane away from the epicenter. The orientation of the fault plane is specified by its strike angle and dip angle. The slip or offset between the fault surfaces can have any orientation in the plane. These and other parameters can be determined from recordings of the seismic waves. The extent of the fault surface is usually determined by the location of aftershock hypocenters. The hypocenter can be located anywhere on the fault fracture

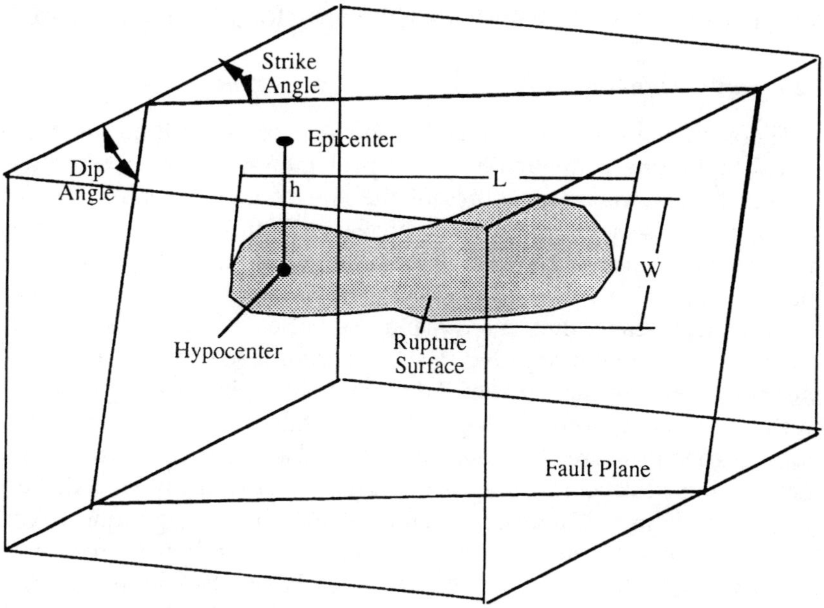

FIGURE 2-3. Idealized Model of an Earthquake Source.

Source: Based on D.M. Boore. 1977. "The Motion of the Ground in Earthquake," *Scientific American* 237(6, December): 68–78.

surface. The length, *L*, and width, *W*, of the rupture surface are shown on the figure. In this illustration, the fault rupture did not reach the earth's surface.

The average depth of most earthquakes is 5 km to 200 km. The maximum offset along the fault trace in the 1906 San Francisco earthquake was 6 m and that along the 1992 Landers earthquake was 7.5 m. The duration of shaking is determined by the length of the rupture surface, the location of the hypocenter on the fault surface, and the site soil conditions. For a given length of rupture surface, the duration of shaking is shorter if the hypocenter is located near the center of the rupture surface than if it is located near the end of the rupture surface. In an earthquake with the hypocenter located near one end of the rupture surface, a site located near the other end of the rupture surface will generally be more severely shaken than a site near the originating end. In general, shallow earthquakes are more damaging than deep earthquakes because seismic energy for a deep earthquake is attenuated before it gets to the surface. Soil conditions at a site can also significantly influence the ground motion. Thus, there can be large variations in the severity of shaking at a site for a given size earthquake for different source parameters and site conditions.

2.2 QUANTIFYING THE SIZE AND INTENSITY OF EARTHQUAKES

2.2.1 Earthquake Size

There are several ways in which the size of an earthquake can be described. Engineers frequently use the peak ground acceleration recorded during the earthquake to represent the size of an earthquake. Because strong motion seismographs record acceleration, this number is often the most readily available. Unfortunately, peak ground acceleration is a poor measure of the size of an earthquake, in part because the limited number of instruments means that many larger ground accelerations probably occurred but were not recorded. The peak ground acceleration is only one measure of the size of an earthquake, and it only applies at a single location. Even from an engineering perspective, the peak ground acceleration may not be a good measure of the damage potential at the location where it was measured, as peak velocity has a better correlation to some types of damage.

The size of an earthquake is usually specified by its magnitude. Several measures of magnitude have been devised. The most broadly recognized magnitude scale was developed by Richter and Gutenberg in the 1930s and 1940s. As the understanding of earthquakes improved and earthquake recording devices improved and became more widely distributed, other magnitude scales were developed, such as Local Magnitude, Surface Wave Magnitude, Body Wave Magnitude, Stress Drop Magnitude and Moment Magnitude (see Section 2.7.3). All of the scales are logarithmic in character, so that an increase in magnitude of one corresponds to about a 30-fold increase in the amount of energy released. At this time, the U.S. Geological Survey uses the Moment Magnitude, M_w, to characterize earthquake size. This scale was proposed in 1979[2] and uses seismograph records, the spatial distribution of intensity data, and ground deformations (geodetic data) to determine the magnitude. Several of the other magnitude scales tend to underestimate the size of large earthquakes, but it seems that the Moment Magnitude has the best overall correlation to energy released in an earthquake. Table 2-1 gives the Moment Magnitude of several historic earthquakes.

Descriptive names, such as "great" or "moderate," are often used to characterize the size of earthquakes. This document will use the descriptive terms used by the USGS National Earthquake Information Center. Table 2-2 shows the number of different size earthquakes based on data gathered since 1900.

The local damage potential, as measured by the severity and duration of the ground motion and its frequency content, is greatly influenced by local geological and site soil conditions. Focusing effects resulting from certain geological conditions have been observed and site soil conditions can significantly change the amplitude and frequency content of ground motions. The peak ground motion at a site near a fault rupture increases with the magnitude of the earthquake. However, for distances within a mile or two of

TABLE 2-1. Magnitudes and Duration of Shaking for Notable Earthquakes.

Earthquake	Moment Magnitude (M_W)	Duration of Strong Shaking
1811 New Madrid, MO (December 16)	8.1	—
1812 New Madrid, MO (January 23)	7.8	—
1812 New Madrid, MO (February 7)	8.0	—
1886 Charleston, SC	7.0	—
1906 San Francisco, CA	7.8	—
1964 Alaska	9.2	6 minutes
1971 San Fernando, CA	6.6	8–10 seconds
1985 Loma Prieta, CA	6.9	5–12 seconds
1995 Northridge, CA	6.7	10–12 seconds

TABLE 2-2. Frequency of Occurrence of Earthquakes of Different Magnitudes.

Descriptor	Magnitude	Number/Year (average)
Great	8.0 and higher	1
Major	7.0–7.9	18
Strong	6.0–6.9	120
Moderate	5.0–5.9	800
Light	4.0–4.9	6,200 (estimate)
Minor	3.0–3.9	49,000 (estimate)

Source: USGS National Earthquake Information Center web site, Geologic Hazards, URL http://www.neic.cr.usgs.gov/neis/eqlists.html/

the fault rupture, increases in magnitude above 6.5 do not cause a significant increase in peak ground motion. The area of severe shaking and its duration will continue to increase as the magnitude increases.

2.2.2 Earthquake Intensity

The severity of an earthquake, as measured by the type and amount of damage it causes, is usually described by its intensity. On a worldwide basis, one of four intensity scales is used to describe the local severity of an earthquake. In the United States, the Modified Mercalli Intensity (MMI) scale is usually used. Each earthquake has a distribution of intensities determined by the local damage. The MMI scale ranges from I to XII. For example, the toppling of chimneys is associated with an intensity VII, and heavy damage and some collapse of ordinary masonry structures is associated with an intensity

IX. Appendix A describes damage associated with the MMI scale. Some power system damage does not correlate well with the MMI scale. For example, severe substation damage has been observed at MMI levels as low as IV.

A map that shows contours of areas of equal intensity is referred to as an *isoseismal* map.

The MMI scale, in addition to providing a measure for earthquake damage, is a vital tool for estimating earthquake magnitude for earthquakes that predate instrumental data.

2.3 EFFECTS OF EARTHQUAKES

Knowledge of the effects of earthquakes on power systems has been gained from observing earthquake damage that has occurred in many parts of the world. Each of the following sections will discuss one of the major effects of earthquakes: ground vibration, soil liquefaction, soil-structure interaction, earthquake-induced landslides, subsidence, ground faulting, and earthquake-induced water waves. There is an emphasis on those effects that are most important to power systems.

2.3.1 Ground Vibration

When an earthquake occurs, seismic energy radiates away from the fault rupture in the form of ground vibrations that induce vibration in the structures and equipment resting on the ground. In general, the severity of the ground shaking decreases as the distance from the source increases. However, local soil conditions can significantly change the character of the ground motion. As the depth of soft soil increases at a site, the low-frequency component of the ground motion is amplified, the high-frequency component tends to be attenuated, and the duration of strong shaking is increased. Ground vibration levels may be amplified by a factor of three or more because of site conditions. Figure 2-4 shows the effect of soft ground conditions on ground motion compared to that of a rock site.

Earthquake excitations can be characterized by the amplitude of the shaking, its frequency content, and its duration. The frequency of the high-energy content of earthquake ground motions often coincides with the natural frequencies of power system equipment. Ground-induced vibration is the major cause of power system damage.

The response of ground-supported structures and equipment is determined not only by the amplitude of the ground motion but also by whether the frequency content of the ground motion matches the natural frequencies of the items being excited. Likewise, the response of equipment mounted in structures is influenced by the coincidence of a natural frequency of the structure and of a natural frequency of the equipment. Thus,

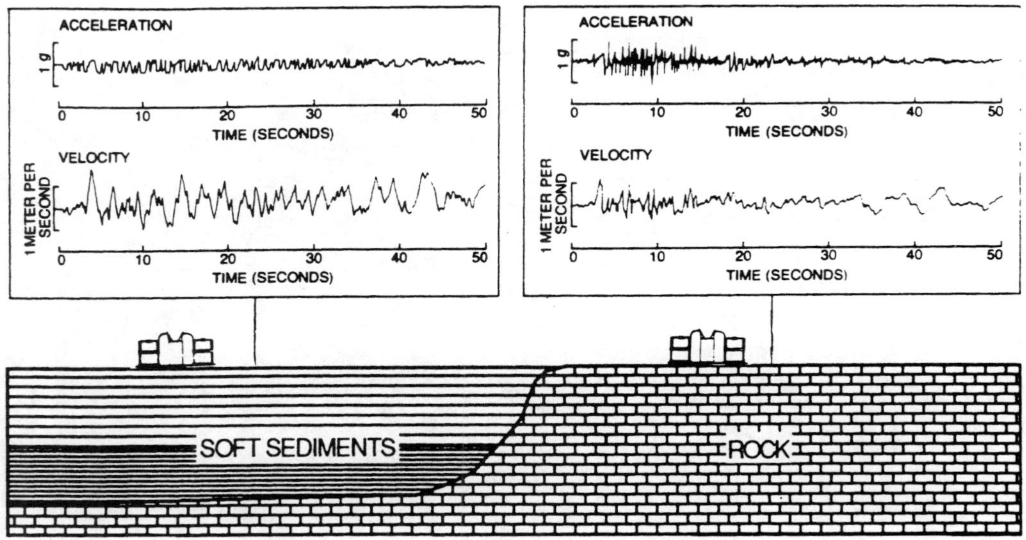

FIGURE 2-4. Local Site Soil Conditions Affect Ground Motion.

Source: Adapted from D.M. Boore. 1977. "The Motion of the Ground in Earthquake," *Scientific American* 237(6, December): 68–78.

equipment mounted in structures may experience a dynamic response much larger than if mounted directly on the ground. The approaches suggested in this guide do not involve dynamic analysis of equipment but use simplified methods.

Earthquake-induced vibrations have caused massive failures of porcelain members used in equipment (Figure 2-5), have damaged structures (Figure 2-6), and have caused poorly anchored transformers to slide from the supports and tip over (Figure 2-7).

Outside of California, there are very few strong-motion earthquake records from significant earthquakes. Many of the records obtained in California show that ground motion can be sensitive to site conditions. Utilities with important facilities for which soil conditions may influence seismic exposure at the site should maintain a strong-motion instrument at the site. While this will not prevent damage should an earthquake occur, it would provide valuable information for reconstruction and planning.

2.3.2 Soil Liquefaction

Granular, water-saturated, low-cohesive soils can lose their shear strength and liquefy when they experience vibrations. Liquefied soil has been observed to flow on 1% grades. Surface-supported structures have settled several feet below grade, and buried tanks have floated to the surface.

FIGURE 2-5. Circuit Breakers Were Severely Damaged Because of Ground Vibrations.

FIGURE 2-6. Capacitor Racks Failed and Collapsed When Subjected to an Earthquake.

FIGURE 2-7. A Rail-Supported Transformer Was Poorly Anchored and Tipped Over.

When soil below the ground surface liquefies, the surface soil can break into blocks that can move back and forth or can shift laterally in one direction, a condition called *lateral spreading*.

Factors contributing to liquefaction include the amplitude and duration of shaking, height of the water table, the soil density, and the granular character of the soil. These conditions are common adjacent to rivers and lakes, where power-generating facilities are frequently located. Figure 2-8 shows a bridge in which one of the piers sank below the water surface because of liquefaction, dropping two spans into the water. The banks of rivers and the shores of lakes tend to move towards the water when soils below the surface liquefy. Figure 2-9 shows damaged cable ducts at a substation located adjacent to a river. There are cases where power distribution poles have settled almost 2 m when soil liquefied. The widespread presence of high water tables in the Eastern United States suggests that more liquefaction can be expected there than in California.

2.3.3 Soil–Structure Interaction

Large structural support loads on soft soils can cause a rocking motion due to the interaction of the structure with the soil that supports it. This effect has been observed on massive transformers and transmission line support towers. Although this has been observed when soils liquefy, it can also occur without liquefaction. In the Northridge earthquake, bus connec-

FIGURE 2-8. A Pier in the River That Supported a Bridge Sank below the Surface When the Soil Liquefied.

FIGURE 2-9. Ducts for Control Cables at a Substation Adjacent to a River Were Damaged Because of Lateral Spreading.

tions to several transformers were damaged because of the rocking of the transformers. Figure 2-10 shows permanent tilt of a transformer pad relative to an adjacent concrete walk.

Appendix B discusses results from an unpublished study of soil-structure interaction during and soil remediation after the Northridge earthquake.

2.3.4 Earthquake-Induced Landslides

There are many regions in which earthquake-induced shaking triggers landslides. The topography and soil conditions are the primary control variables. Should the earthquake occur during a rainy season when soils are saturated, landslide potential increases. Slides and rock fall can cause excessive deformations in the ground, and the falling material could sweep away

FIGURE 2-10. Rocking of a Transformer Due to Soil-Structure Interaction Can Cause Large Movement of Conductors Connected to the Transformer Bushings.

structures and equipment in its path. However, there is no record of a landslide causing such damage to a major substation or a transmission tower. In the Northridge earthquake, several transmission towers located on ridges or edges of steep slopes had foundation failures. While some of these failures may have been associated with ridge shattering (the focusing of seismic energy at ridges and peaks), others appeared to be caused by soil failure in a landslide, Figure 2-11.

2.3.5 Subsidence

Under certain conditions, an earthquake may cause the ground to settle. Inadequately compacted fills are vulnerable. Subsidence can cause severe damage to buried utilities such as water, gas, and oil lines. The approach to a bridge abutment is one of the most frequently observed places of subsidence, Figure 2-12. Figure 2-13 shows soil at a gas metering station of a power-generating station; the soil settled about 15 cm, leaving the pipe unsupported, but the gas lines were undamaged.

2.3.6 Ground Faulting

Fault surfaces of shallow earthquakes frequently extend to the ground surface. Along much of the San Andreas fault, the offset after an earthquake

FIGURE 2-11. The Foundations of Some Transmission Line Towers Located at the Edge of Steep Slopes Failed When Soil Moved in a Landslide.

FIGURE 2-12. Subsidence of the Road Bed at Bridge Abutments Is Frequently Observed after Earthquakes.

FIGURE 2-13. Soil around the Gas Lines at a Power-Generating Plant Metering Station Subsided about 15 cm.

is limited to a horizontal displacement along the fault. Frequently, there is also a vertical offset. In areas with multiple fractures, the area is referred to as a *fault zone*. Anything spanning the fault, such as buried pipe or cable or a structure, can experience severe deformations when significant offsets occur. Offsets at faults can be quite large; the maximum associated with the 1906 San Francisco earthquake was 6 m. The 1960 Chile earthquake (M_W = 9.5) had an average fault offset of 20 m. In some earthquakes, particularly in the eastern United States where there are thick alluvial deposits, the faulting may not extend to the ground surface and thus might not be observed.

Long continuous structures, such as pipelines or buried transmission cables, are more likely to be affected by faulting than isolated structures. Faulting at concentrated facilities, such as substations, has been rare. In the 1992 Landers earthquake, a fault trace passed under a transmission tower between footings so that two footings were on each side of the fault. However, even though the fault offset at the tower was almost 3 m, the tower continued to support the transmission lines (Figure 2-14).

2.3.7 Earthquake-Induced Water Waves

Large earthquakes that occur offshore and have vertical components can generate long-period water waves called *tsunamis* (see also Section 2.7.5). Typically, these waves are barely perceptible in deep water; however, they can become massive waves when encountering a land mass. A tsunami travels great distances at speeds of about 800 km/h with little attenuation. As a

FIGURE 2-14. A Fault with an Offset of almost 3 m Passed between the Legs of a Transmission Tower without Causing It To Fail.

Note: Also see Figure 6.1.

result, waves generated thousands of kilometers away can wreak havoc when they reach a distant shore. For example, in 1964, Crescent City, California, was devastated by a tsunami generated in Alaska at a distance of more than 2,000 km. The loss of life in Crescent City was larger than that which occurred in Anchorage because of the earthquake. The 1993 Hokkaido-Nansei-Oki, Japan, earthquake ($M_W = 7.8$) created a tsunami that had a 15 to 20 m vertical runup along the coast of Okushiri Island, located 48 km from the epicenter. The topography created by a river canyon caused the water to rise 31 m above sea level. The action of the wave stripped clean a portion of the town of Aonae (Figure 2-15). Because of the destructive power of tsunamis, power-generating stations located on coasts may be vulnerable.

2.4 REGIONAL DIFFERENCES IN EARTHQUAKES AND ASSOCIATED HAZARDS

In the last 50 years, the understanding of earthquakes has been revolutionized by the discovery of plate tectonics. While this has provided a framework for understanding why earthquakes occur in many regions of the country, the short historical record in the United States does not provide sufficient data for a clear understanding of seismic hazards. Along a few West Coast faults, the earthquake history for the last several thousand years has been uncovered using trenching and carbon-dating methods. Using these methods, the number, the time interval between events, and the magni-

FIGURE 2-15. Part of the Town of Aonae, Japan, Was Washed Away When It Was Hit by a Tsunami.

tudes of large earthquakes have been used to estimate probabilities of future events. However, this method can be used only when special conditions exist, and this data is not available for most faults. Further, little such data is available outside of California. In many regions, such as the eastern United States, basic mechanisms are less well understood, and much remains to be learned. Thus, the uncertainties about location, size, and frequency of occurrence are much larger in the Midwestern and Eastern sections of the United States. Even in the regions that are best understood, estimates are based on relatively short observation intervals. Observations made in China, which has a record dating back 5,000 years, show a shift in seismic patterns over thousands of years. Even with this long record of past earthquakes, the disastrous Tangshan earthquake of 1978, in which more than 250,000 people died, occurred in a region rated as having low to moderate seismicity.

A map showing the areas of damage resulting from four historic earthquakes is enlightening because it demonstrates that the attenuation in the eastern United States is much less than in California (Figure 2-16). The New Madrid earthquakes of 1811 and 1812 impacted a much larger area than the 1906 San Francisco Earthquake. The attenuation of seismic waves in the Midwest is much less than in California, so that even a strong Eastern earthquake can affect a larger area than a great Western earthquake. This map also illustrates that large earthquakes have occurred outside of California. Indeed, many people are surprised to learn that the largest and potentially most damaging earthquakes in the United States had their origin in the Midwest, not in California.

The effect of an earthquake on power facilities is further influenced by the local geology and site conditions. In general, conditions in the Midwest and East lend themselves to more damaging earthquakes because large areas have a high potential for soil liquefaction and many sites are covered with soft soils that can amplify ground motions.

2.5 REGIONAL SEISMICITY OF THE UNITED STATES

To review the seismicity of the United States, the country has been divided into three regions: Western, Central, and Eastern. The earthquake history of each region will be reviewed. A map for each region shows the epicenters using a symbol that identifies the peak MMI generated by the earthquake. For each region, selected isoseismal maps from historic earthquakes have been drawn. Each earthquake had a peak MMI value of at least VI. Severe power system damage has occurred with MMI values less than IV. The intent is to illustrate that all parts of the country have experienced potentially damaging earthquakes. Many of the isoseismal maps are from earthquakes that predate modern power systems; however, they provide a historical basis for damaging earthquakes. For each region, a table gives by

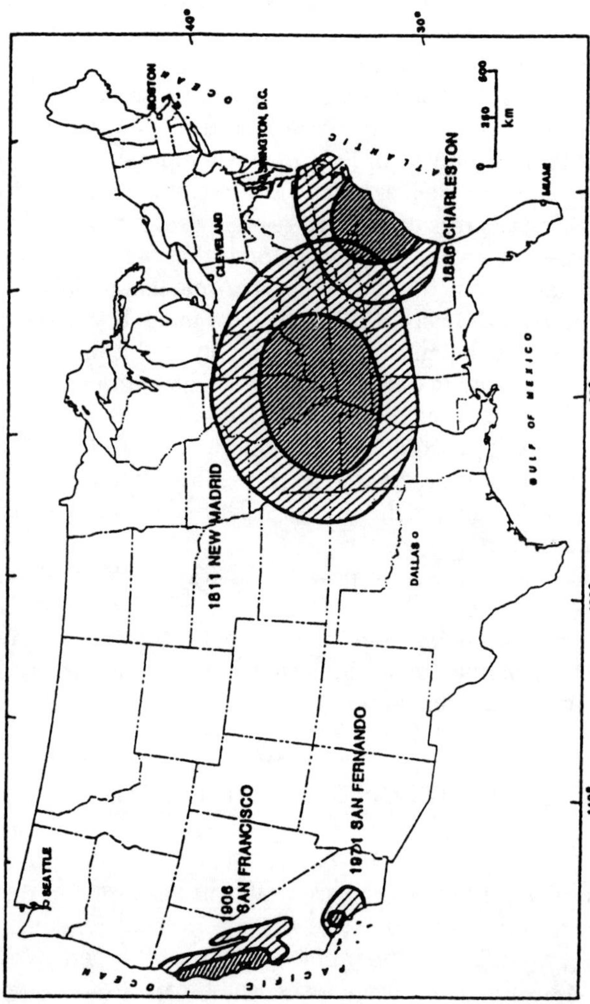

FIGURE 2-16. Areas Impacted by Four Historical Earthquakes.

state the number of earthquakes with a peak MMI for any earthquake above V or IV. The most significant earthquakes in the region and other factors that influence the seismicity of the region will be briefly discussed. The peak MMI levels for earthquakes with epicenters in each state are listed to indicate that most states have experienced earthquakes with a MMI of VI or larger. All but three states (Iowa, Wisconsin, and Maryland) have had qualifying earthquakes.

Using the list of earthquakes that have had epicenters in a state, one can underestimate the seismic exposure for the state. For example, the first earthquake of the New Madrid series occurred on December 16, 1811, in Arkansas. This earthquake generated higher intensities in Illinois, Indiana, Tennessee, Missouri, and Mississippi than any of the earthquakes with epicenters in these states. Most of the data in this section are drawn from a USGS Professional Paper[3]. This publication reviews the seismicity of the United States from 1568 through 1989, and the survey in this document is limited to that time period. Because of the general awareness of the high seismicity in Alaska and Hawaii, these states have not been discussed here.

2.5.1 Western Region

Figure 2-17 shows a map of the Western region with each earthquake location indicated by a symbol identifying its peak MMI. The isoseismals are also shown for the following earthquakes: 1857, Fort Tejon, California; 1872, Owens Valley, California; 1872, central Washington; 1906, San Francisco, California; 1949, Puget Sound, Washington; 1954, Dixie Valley-Fairview Peak, Nevada; 1959, Hebgen Lake, Montana; and 1983, Borah Peak, Idaho. Table 2-3 lists the number of events originating in each state with a given peak MMI.

TABLE 2-3. Seismicity in States in the Western Region.

	Number of Events at MMI Levels								
State	No.[a]	XI	X	IX	VIII	VII	VI	V	IV
California	1500	2	3	13	47	125	334	313[b]	—[b]
Idaho	43			1	1	4	12	8	5
Montana	67		1		4	4	35	16	3
Nevada	320		4	2	2	10	24	43	42
Oregon	82					1	10	4	3
Washington	122			1	3	6	38	13	
Total	2134	2	8	17	57	150	453	397	53

[a] Total number of events in the earthquake catalog from 1568 through 1989.

[b] In California MMI V and IV were excluded unless the magnitude was above 5.5.

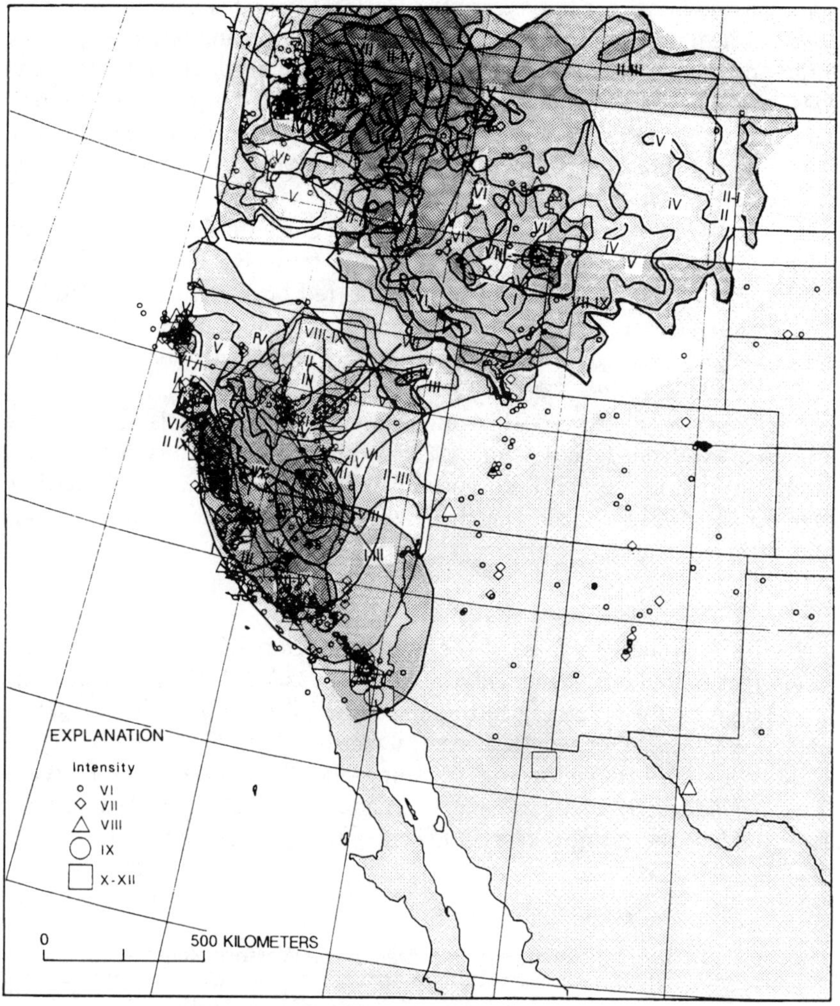

FIGURE 2-17. Epicenters and Isoseismals for Selected Earthquakes in the Western Region.

Of the three selected regions, the Western region is the most seismically active. In California, the period between the 1933 Long Beach and the 1971 San Fernando earthquakes is generally considered quiescent, because there was no significant damage to power system facilities or to the community at large; however, the earthquake catalog has more than 200 entries during this supposedly quiet period. Starting with the 1971 San Fernando earthquake, there have been 10 California earthquakes that have caused damage to power system facilities. This apparent increase in seismic activity may be related to the introduction of more high-voltage equipment, which is more vulnerable to earthquake damage, or to the unfortunate location of the epicenters. Some of the earthquakes that damaged power system facilities were relatively small. For example, the 1988 Tejon Ranch earthquake (Local Magnitude, M_L, = 5.4) had a peak MMI of V; however, it damaged 10 of 18 phases of 230-kilovolt (kV) circuit breakers at the site. The intensity at the switchyard site is not known; a strong motion instrument in the basement of a nearby water-pumping station had a peak ground acceleration of only 0.08*g*.

The earthquakes that take place along the west coast of California are typically associated with movements along the boundary between the Pacific and North American plates. These plates move relative to each other at a rate of about 2 cm/yr. In some locations, like Southern California, the North American Plate (which includes most of the United States and Canada) is moving horizontally in a southerly direction relative to the Pacific Plate (which makes up much of the floor of the Pacific Ocean). This region is not only the most seismically active in the contiguous United States but also the best understood in terms of source mechanisms and the character of expected ground motions. A review of Figure 2-17 shows that there were five events with a peak MMI of X to XII along the San Andreas fault. There were also five events with a peak MMI of X to XII in a line from Lone Pine in the Owens Valley, California, to Winnemucca, Nevada, so that large earthquakes in the Western region are not limited to the San Andreas. Much of California and Nevada are desert, so water tables are low and there is a low potential for soil liquefaction.

Recently paleontologic earthquake investigation has identified large subduction earthquakes (M_w above 9) along the coast north of Eureka, California, up to British Columbia, Canada. This discovery has significantly changed the seismic risk maps used by the building codes in this region.

2.5.2 Central Region

Figure 2-18 shows a map of the Central region with each earthquake location indicated by a symbol identifying its peak MMI. The isoseismals are also shown for the following earthquakes: 1811, Arkansas (1st New Madrid); 1882, Colorado; 1884, Lima, Ohio; 1887, Sonora, Mexico; 1891, southern Illinois; 1895, Charleston, Missouri; 1909, Wabash River valley, Indiana; 1925, Texas Panhandle; 1934, Kosmo (Hansel Valley), Utah; 1964, northwest Nebraska; 1980, northern Kentucky; and 1984, eastern Wyoming. Table 2-4 lists the number of events originating in each state with a given peak MMI.

2.5.2.1 New Madrid Earthquake Series. One of the most severe series of earthquakes within the contiguous United States, as measured by the area subjected to severe shaking, is referred to as the New Madrid earthquakes. Severe shocks started in December of 1811 and continued over a period of several months. There were three major events with M_w estimated at 8.1,

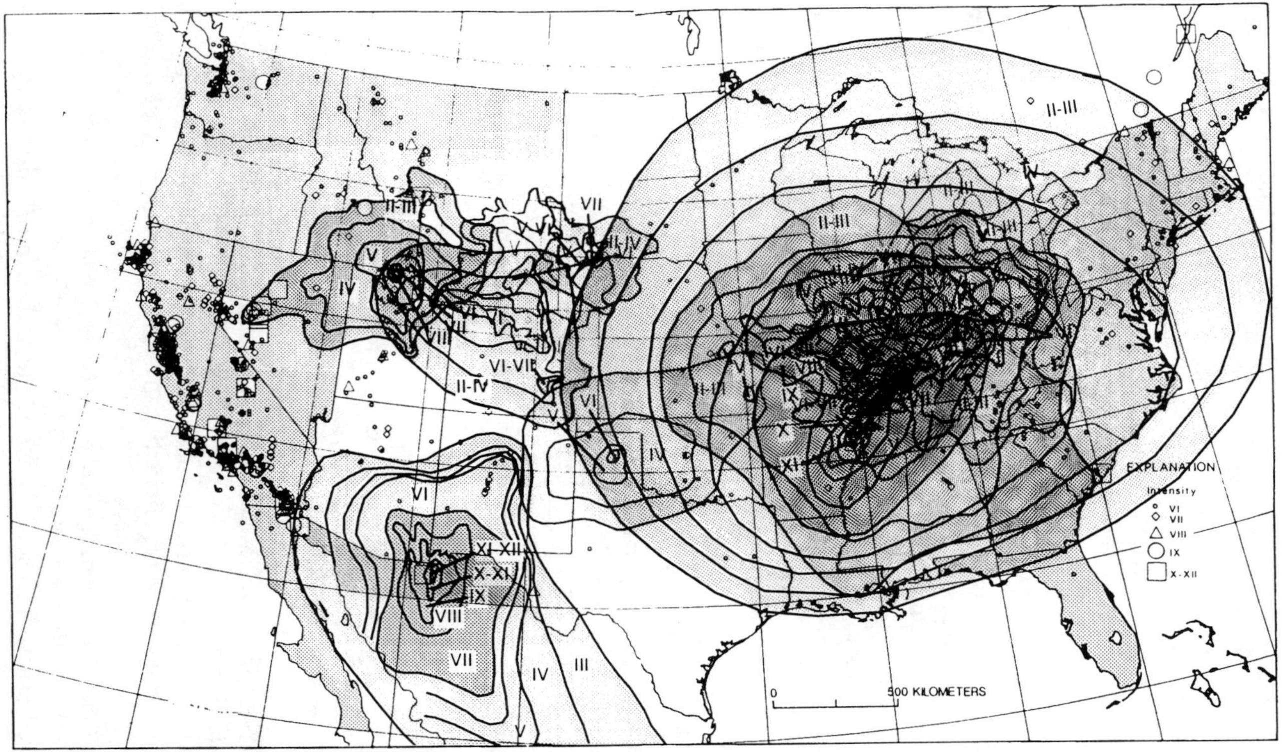

FIGURE 2-18. *Epicenters and Isoseismals for Selected Earthquakes in the Central Region.*

TABLE 2-4. Seismicity in States in the Central Region.

	Number of Events at MMI Levels								
State	No.[a]	XII	XI	X	IX	VIII	VII	VI	V
Alabama	6						1	5	
Arizona[b]	17	1					3	12	2
Arkansas	16		1	1			3	8	3
Colorado	21						2	19	
Illinois	31						12	18	1
Indiana	9						2	5	2
Kansas	6						2	4	
Kentucky	11						1	8	2
Louisiana	1							1	
Michigan	3					1	1	1	
Minnesota	3							3	
Mississippi	2							2	
Missouri	22	1	1			1	2	14	3
Nebraska	6						2	4	
New Mexico	37						7	29	1
North Dakota	1							1	
Ohio	15					1	5	9	
Oklahoma	11						2	9	
South Dakota	7							6	1
Tennessee	16						2	12	2
Texas	10					1		7	2
Utah	54					5	8	31	10
Wyoming	20						1	8	11
Total	325	2	2	1		9	56	216	40

[a] Total number of events in the earthquake catalog from 1568 through 1989.

[b] Epicenter just south of the Arizona–Mexico border.

7.8, and 8.0, and epicentral intensities of X to XII. In addition, there were 5 aftershocks with intensity X, 10 with intensity IX, and 40 with intensity VIII.[4]

The cause of the New Madrid earthquakes is beginning to emerge. The relative motion of locked plates causes stress to develop throughout the plates. Recently, what appears to be an ancient rift has been discovered, which roughly parallels the Mississippi River. The long dimensions of this feature account for the possibility that such large earthquakes could occur in the midcontinent. However, the length of fault-releasing energy in the mid-continent is expected to be significantly shorter than for a comparable earth-quake in the West, because of the character of the basement rock. The existence of these midcontinent stress fields also means that strong earthquakes (M_w about 6.5) can occur almost anywhere in the United States.

The attenuation characteristics of the midcontinent earthquakes influence their impact in many ways. First, as indicated in Figure 2-16, for a given magnitude earthquake, an earthquake in the Midwest will induce severe shaking in a much larger area than one of similar magnitude in the Western region. Thus, cities separated by moderate distances may be affected by the earthquake and be unable to provide assistance or mutual aid as would be the case in the West. Second, more substations will be impacted, so that the robustness of the system provided by redundancy may be overwhelmed. Also, those sites further away from the epicenter that experience fairly severe shaking will experience shaking that will last longer and will have a greater low-frequency content. High-voltage substation equipment has low natural frequencies, and longer duration shaking would cause more damage than would be experienced by a similar magnitude earthquake in the West.

More severe and widespread liquefaction can be expected in the Midwest as compared to the West. Liquefaction is sensitive to the duration of shaking, so even if everything else were the same, more problems would be expected. However, an additional factor makes liquefaction an even more serious problem in the Midwest. The most important factor contributing to liquefaction is the presence of water-saturated, coarse-grained soils. These conditions are much more prevalent in the Midwest than in the West. Observations made at the time of the New Madrid earthquakes attest to this, and large sand boils caused by liquefaction remaining from this large event can still be seen. Liquefaction was observed as far as 500 km from the epicenter.

In most of the Eastern and Midwestern United States, the basement rock that fractures during an earthquake is covered with alluvial deposits, often to considerable depth. As a result, very little surface faulting can be expected when an earthquake occurs. However, large areas of subsidence and differential settling can be expected.

In the West, the high population density regions that are subjected to earthquakes lie primarily along the coast, where temperatures are moderate. Should a catastrophic Midwestern or Eastern earthquake occur in winter when temperatures are low, the impact of the earthquake may be more severe for two reasons. First, the damage experienced by high-voltage substation equipment may be more severe. The most vulnerable members in substations are porcelain insulators and bushings. Tests have shown that the material often used to bond the porcelain to the end flanges is temperature sensitive. At low temperatures, this material becomes less plastic. Thus, the damping exhibited by these members can drop significantly, and a larger dynamic response can be expected with a commensurate increase in failures. Second, should the earthquake coincide with a winter storm, the problems associated with transportation and field repairs would be severe, and recovery times would be much longer than those experienced in California.

2.5.3 Eastern Region

Figure 2-19 shows a map of the Eastern region with each earthquake location indicated by a symbol identifying its peak MMI. The isoseismals are also shown for the following earthquakes: 1811, Arkansas (New Madrid); 1884,

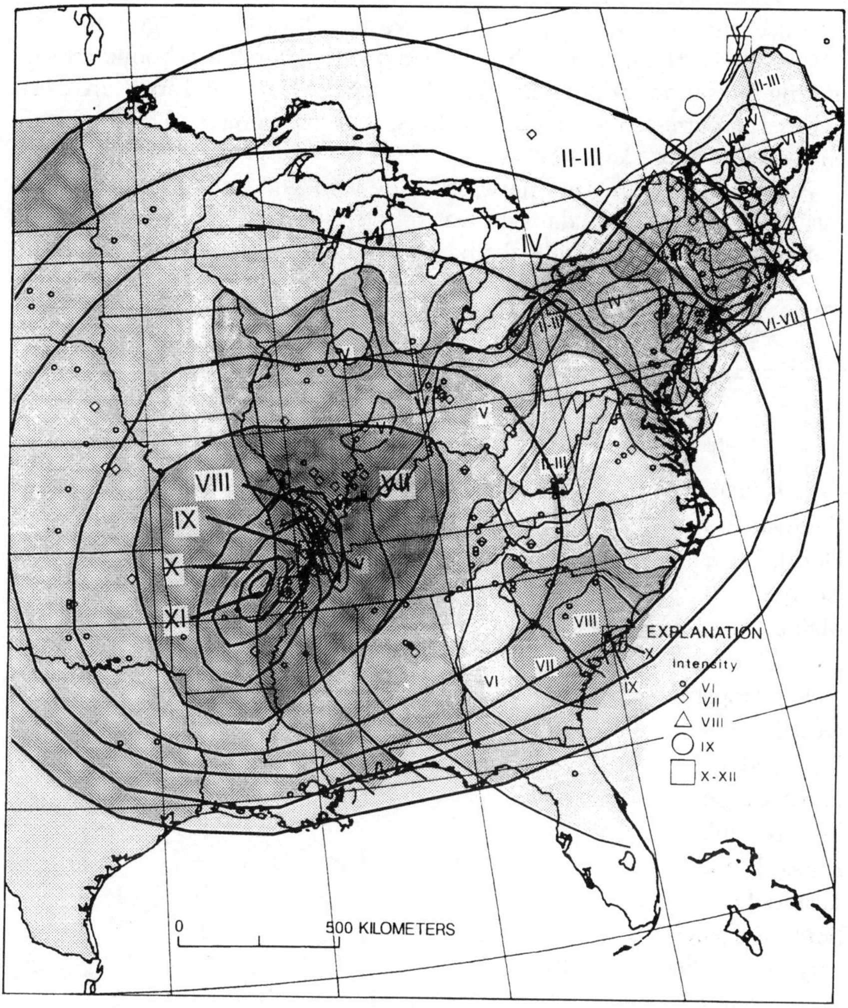

FIGURE 2-19. Epicenters and Isoseismals for Selected Earthquakes in the Eastern Region.

New York; 1886, Charleston, South Carolina; 1940, New Hampshire; and 1944, Massena, New York. Table 2-5 lists the number of events originating in each state with a given peak MMI.

Like the Central region, the East is expected to be more severely impacted than the West because of the greater potential for soil liquefaction and lower attenuation of seismic energy.

2.5.3.1 Charleston, South Carolina. In 1886, an earthquake with magnitude estimated at M_w 7.0 was centered 40 km northwest of Charleston. Recent earthquakes have had depths of 3 to 13 km, suggesting that the 1886 event also occurred in the upper crust. There are clues that suggest that the Charleston earthquakes may be associated with a horizontal boundary separating formations of different ages. This type of horizontal structure exists over large parts of the East Coast; thus, there is a potential of significant earthquakes over a larger area.

Almost all structures in Charleston were damaged, and few escaped serious damage. Structural damage was reported several hundred kilometers from Charleston in central Alabama, central Ohio, eastern Kentucky, south-

TABLE 2-5. Seismicity in States in the Eastern Region.

	Number of Events at MMI Levels						
State	No.[a]	X	IX	VIII	VII	VI	V
Connecticut	3				1	2	
Delaware	1				1		
Florida	2					2	
Georgia	6					5	1
Maine	10				2	7	1
Massachusetts	13	1	1	1	1	8	1
New Hampshire	10		1		2	7	
New Jersey	10				1	5	
New York	24			2	6	16	
North Carolina	7			2	6	16	
Pennsylvania	8				1	7	
Rhode Island	1					1	
Vermont	1					1	
South Carolina	22	1			2	17	2
Virginia	19			1	1	12	5
West Virginia	1					1	
Total	134	2	2	4	20	96	10

[a] Total number of events in the earthquake catalog from 1568 through 1989.

ern Virginia, and western West Virginia. The southern part of Lake Michigan, including Chicago, experienced a MMI of V associated with this earthquake. A notable feature of the earthquake was the development of large cracks in the ground that paralleled rivers, due to liquefaction and lateral spreading.

2.5.3.2 Cape Ann, Massachusetts. The 1755 Cape Ann earthquake had its epicenter offshore in the North Atlantic. It had a peak intensity MMI of VIII and affected the area around Cape Ann and Boston. The earthquake was felt from Halifax, Nova Scotia, south to the Chesapeake Bay in Maryland, and to Lake George, New York. This earthquake created a 2-m tsunami in the West Indies. The source of the earthquake and the persistent seismicity are not well understood. There are certain geological features in the area, including an ancient rift and a transform fault, that could account for the earthquakes.

2.6 SUMMARY OF DIFFERENCES BETWEEN EARTHQUAKES IN CALIFORNIA AND OTHER REGIONS

- Similar peak accelerations can be expected for similar magnitude earthquakes throughout the United States.
- A single large event will impact a much larger area outside of the West Coast.
- More extensive areas of soil liquefaction and subsidence can be expected in the Central and Eastern regions of the United States as compared to the Western region.
- In the Central region, a series of major events and many significant aftershocks can be expected over a period of several months, as compared to the West, if the New Madrid earthquakes are representative.

Outside of California, which recently introduced seismic design codes, damage to the infrastructure needed during postearthquake recovery, such as the transportation system, may impede the recovery of power facilities.

2.7 COMMONLY USED TERMS

2.7.1 Fault and Fault Trace

A *fault* is the fracture surface within the earth along which the earth's crust moves relative to adjacent parts of the crust. This surface forms a weak section so that future earthquakes will tend to occur along the same fracture. The intersection of the fault plane with the surface of the earth is called the *fault trace*. In some earthquakes, the fault trace can be seen in the form of an offset at the ground surface. In some regions, the fault is not a well defined surface but is characterized by many parallel fractures and is referred to as a *fault zone*.

2.7.2 Hypocenter and Epicenter

The *hypocenter* is the location on the fault where an earthquake starts. The *epicenter* is a point on the earth's surface directly above the hypocenter.

2.7.3 Earthquake Magnitudes

Several different magnitude scales are commonly used to characterize the size of an earthquake. They are derived from seismograms obtained from an earthquake and are related to the energy released in the earthquake. All of the scales are exponential in character so that an increase in magnitude of one corresponds to about a 30-fold increase in the amount of energy released. The U.S. Geological Survey now uses the Moment Magnitude (M_w) to characterize the size of earthquakes. Most other magnitude scales tend to underestimate the size of large earthquakes. Other magnitude scales that are used include the Richter (M), Local (M_L), Surface Wave (M_s), and Body Wave (M_b).

2.7.4 Intensity Scales

Earthquake intensity levels are used to describe the local impact of an earthquake based on observed effects. Both natural effects, such as soil liquefaction, and the effect on structures are used to characterize the intensity. Several different intensity scales are used throughout the world, but the most frequently used in the United States is the Modified Mercalli Intensity Scale (MMI). The definition of the scale is given in Appendix A.

2.7.5 Tsunamis

Tsunami is the term used to describe an earthquake-generated tidal wave. Earthquakes with significant vertical offset on ocean bottoms can generate large waves. They are often associated with earthquakes along subduction zones. These waves are barley noticeable in deep water, where they may be only 30 cm high. They can travel thousands of kilometers at velocities as high as 800 km/h. When it impinges on a coast line, the tsunami can rise as high a 15 m above sea level. Certain topographical features, such as a river canyon, can extend the wave height to as high as 30 m. Another potential source of large earthquake-generated waves is a large landslide that occurs below water or that enters the water.

ENDNOTES

1. D.M. Boore. 1977. "The Motion of the Ground in Earthquake," *Scientific American* 237(6, December): 68–78.
2. T. Hanks and H. Kanamori. 1979. "A Moment Magnitude Scale," *Journal of Geophysical Research*, 84(B5, May 10): 2348–2350.
3. C.W. Stover and J.L. Coffman. 1993. *Seismicity of the United States, 1568–1989*. USGS Professional Paper 1527.
4. J.L. Penick Jr. 1981. *The New Madrid Earthquake* (rev. ed.). Columbia, Missouri: University of Missouri Press.

Chapter 3

OVERVIEW OF EARTHQUAKE PERFORMANCE OF POWER SYSTEMS AND FACILITIES

In this section, power systems have been grouped into three types of facilities: transmission and distribution facilities, power-generating stations, and control and communications facilities. Facilities that do not immediately impact system operations, such as design offices and maintenance facilities, are not discussed here. The overall performance of the entire system and the performance of each of these types of facilities are discussed. A detailed discussion of the performance of individual equipment and facilities is given in subsequent sections.

3.1 OVERALL POWER SYSTEM SEISMIC PERFORMANCE

Before the seismic performance of power facilities within the United States can be properly understood, four factors must be kept in mind.

1. There are no data from a major or great earthquake centered in a modern metropolitan area. The evaluation of system performance is primarily based on several moderate and two strong California earthquakes.
2. Knowledge of earthquakes in California has been evolving since the 1920s. Since 1933, a stronger impetus has been given to the seismic design of power facilities. This has been a slow process as changes in design take a long time to be reflected in most facilities in the field. However, the vast majority of facilities subjected to earthquakes since 1970 have had significantly higher seismic specifications (particularly anchorage of substation equipment) than most such facilities outside of California. While other regions may have other more severe design specifications, such as those for high wind loads, the requirements typically have little effect on improving the seismic performance of substation equipment.

3. The moderate magnitude of most recent earthquakes and the high attenuation of seismic energy in California, as compared to most of the eastern United States, mean that in California relatively small areas have been exposed to damaging ground motions. As a result, the damage to power systems in most earthquakes has been confined to one or two facilities. In the eastern United States, equivalent magnitude earthquakes will impact much larger areas, and large areas have a high potential for soil liquefaction. Thus, an eastern earthquake may affect several facilities and be more disruptive to the network, and soil liquefaction at a power facility can be more damaging.
4. Large coal-fueled plants, with their heavy coal storage silos located high in the steam-generation (boiler) structure, have not been put to the test by damaging earthquakes, so their seismic performance is unknown. Within California, all large fossil-fuel power-generating stations are units that burn gas or oil rather than coal. There is no earthquake data available on the performance of coal-fueled generating stations in seismic regions in other countries.

Given these caveats, it can be said that system performance, as measured by power disruption, has been very good. Thus, to date, network redundancy has been adequate to overcome the extensive damage to isolated high-voltage substations. In the case of the Loma Prieta and Northridge earthquakes, where several substations were damaged, the character of the damage and the use of emergency procedures allowed expeditious system restoration.

3.2 POWER TRANSMISSION AND DISTRIBUTION SYSTEMS

The transmission and distribution system can be grouped into three types of elements: transmission lines, distribution lines, and substations.

3.2.1 Transmission Lines

Transmission lines have been very resistant to earthquake damage; their main vulnerabilities are foundation failure of transmission towers or the loss of a tower due to a landslide. Both occurrences are relatively rare. It would appear that the low natural frequencies of lines decouple their mass from the high energy content of earthquakes, and the design for extreme wind, ice, and longitudinal load combinations is adequate for earthquakes.

3.2.2 Distribution Lines

Distribution lines are also seismically robust. Their main vulnerability in the United States is from burn-down, when earthquake induced vibrations cause adjacent lines to come in contact. If they are energized, they will arc and may burn through the line, causing it to fall. Burned-down lines can be a significant source of fires, and they have generated large numbers of calls by the general public to the emergency response system. While repair can be labor intensive, only limited numbers of customers are impacted by any given downed line, and spare parts are usually not needed to effect a repair.

Distribution systems in other countries have had problems with the failure of concrete distribution poles and the failure of houses that supported lines where narrow streets did not allow the use of poles.

3.2.3 Substations

Damage to porcelain members of high-voltage substation equipment has been a recurring problem. Equipment operating at voltages of 115 kV and below performs very well when good seismic installation practices of anchorage and conductor interconnection flexibility are followed. Some types of equipment operating at voltages of 220 kV and above are vulnerable. Generally, the higher the operating voltage, the more vulnerable the equipment. The highest-voltage equipment to be subjected to earthquakes is 500 kV. Little or no equipment operating in the voltage range from between 138 kV and 161 kV is used in California, so its earthquake performance is not known.

Several types of failures are frequently observed. Inadequately anchored rail-supported transformers have fallen from their elevated platforms and have been severely damaged. Leaking or broken bushings are common. Lack of adequate slack in conductors connecting equipment can load and damage bushings and post insulators. Flexible equipment supports have allowed large relative displacements; this tends to aggravate problems with the lack of sufficient slack. Some equipment designs appear to be inherently vulnerable, while other equipment that serves the same function and operates at the same voltage can be quite rugged; for example, some live-tank circuit breakers versus dead-tank circuit breakers. Current transformers and capacitive coupling voltage transformers have been damaged thereby, disrupting system protection. One of the main difficulties when substation equipment is damaged is that there are limited numbers of spare parts or spare replacement equipment. Also, repair and replacement of damaged equipment is a time-consuming task.

3.3 POWER GENERATION FACILITIES

In general, the overall seismic performance of power-generating stations has been good, although coal-fueled plants and large oil- and gas-fueled plants (500 megawatts and above) have had limited exposure. Within California, many generating stations are relatively small and old. Design practices used by California utilities have evolved from 0.1*g* static lateral design force used in the 1920s, to 0.2*g* in the 1950s, and 0.5*g* dynamic design since the early 1970s.

Generating station performance has been good, although some equipment and facilities not causing the plant to shut down have been damaged. Some elements—for example, water and liquid fuel storage tanks—have had mixed performance. Generating stations have been forced off line because of switchyard and substation damage, and they have experienced delays in getting back on line.

Performance of power-generating facilities outside of the United States, primarily in Chile and Japan, has been good. However, a small generating station in Managua, Nicaragua, was severely damaged in 1972.

The discussion here excludes nuclear generating plants, as they are under federal regulation and are designed to much more stringent criteria. Also, because of the special character or limited number of wind, geothermal, and hydropower facilities, they are not discussed in this document.

The switchyards associated with power plants are grouped with transmission and distribution facilities.

3.4 CONTROL, PROTECTION, AND COMMUNICATIONS FACILITIES

In general, control, protection, and communications equipment have performed well. The exceptions are uninterruptible power supplies and emergency power supplies. Inadequate restraint of station batteries at power plants and substations has been a common problem. Protective relays have tripped because of earthquake-induced vibrations. The loss of capacitive coupling voltage transformers has also disrupted or limited system protection. Damage to substation equipment may result in poorer system protection. For example, damage to instrumentation transformers may limit protection to just-over-current conditions and prevent determining the location of faults.

Public Switched Network telephone systems are typically congested, so they cannot be counted on during the emergency response period following an earthquake. Utility-owned communications systems can also become saturated, but the earthquake performance of these facilities has been good. Radio repeaters, which are used to dispatch repair crews, can lose power and stop operating; this impedes the dispatch and control of repair crews.

Chapter 4

APPROACH TO IMPROVED EARTHQUAKE PERFORMANCE

Recommendations

An earthquake preparedness program should be implemented consisting of the following elements:

- hazard and system vulnerability evaluation (4.2),
- earthquake disaster response planning program (4.3), and
- earthquake mitigation program (4.4).

Perform an initial hazard and system vulnerability evaluation to assess the need for an earthquake preparedness program (4.2.1).

Perform a hazard and system vulnerability evaluation to estimate direct losses needed to justify a preparedness program (4.2.2).

Perform a detailed assessment to gather information needed to implement a mitigation program (4.2.2.1).

A disaster response plan should be formulated and include the following elements:

- corporate recovery plan (4.3.2),
- emergency operations center (4.3.4),
- alternate energy control center (4.3.5), and
- periodic exercise of the disaster response plan, and its review and revision (4.3.1).

Earthquake planning should be given high implementation priority (4.3).

Upgrade manuals of practice by incorporating material dealing with the following issues:

- add seismic criteria for site selection and preparation (4.4.2.1);
- incorporate guidance for seismic issues in site layout, including seismic review early in the design process (4.4.2.2);
- incorporate seismic specifications into equipment purchase orders (4.4.2.3);
- develop engineering drawings for anchoring equipment (4.4.2.4);
- develop engineering drawings that assure equipment interconnection flexibility (4.4.2.4); and
- institute a seismic quality assurance check of engineering drawings and as-built field construction (4.4.2.5).

Institute vulnerability assessment of parts of the system that historically have had poor seismic performance (4.4.3).

Diligently apply good practices for renovation of existing facilities and for new construction (4.4.4.1 to 4.4.4.4).

Incorporate seismically hardened power paths in substations to enhance system reliability at reduced cost relative to upgrading the entire switchyard (4.4.4.5).

Develop a plan to upgrade facilities, such as structures, that are not normally renovated (4.4.4.6).

Periodically review and revise the earthquake mitigation program (4.4.5).

It is important to get the support of top management for an earthquake damage mitigation program (4.5.2).

For the program to be effective, mitigation measures must be cost-effective (4.5.3).

Personnel doing design that affects earthquake performance should be familiar with good earthquake practices and take into account effects of earthquakes (4.5.5).

California's experience suggests that power system performance, in terms of disruption, has been good for moderate ($5 \leq M_W < 6$) and strong ($6 \leq M_W < 7$) earthquakes. However, in the recent 1989 Loma Prieta ($M_W = 6.9$) and the 1994 Northridge ($M_W = 6.7$) earthquakes, there was considerable power system damage. The performance of power systems in major ($7 \leq M_W < 8$) and great ($8 \leq M_W$) earthquakes is unknown. The high vulnerability of power system equipment to earthquake damage, the potential for lengthy disruption of service, and the high dependency of modern society on power mandates that the reliability of power systems during and following earthquakes be addressed. In most parts of the country except along the West Coast, the prospect of a large or great earthquake is not imminent. Even in regions of high seismic exposure, such as California, major and great earthquakes do not occur frequently; the last great earthquake in southern California was the 1857 Fort Tejon, and the last great earthquake in northern California was the 1906 San Francisco earthquake.

Power systems in regions outside California where the frequency of earthquakes is lower and seismic awareness is low, may be more vulnerable because of the failure to use good earthquake design practices. In addition, the area affected by a significant earthquake is larger. Thus, in these regions outside of California, measures to improve the earthquake performance should be pursued, although they may proceed at a slower pace than in higher-risk areas and at a cost that is commensurate with the risk. In these regions, the risk of a major earthquake in the next thirty years is relatively low, so that a mitigation program operating at a measured pace is likely to reduce vulnerability before a large earthquake occurs.

4.1 OVERVIEW OF IMPROVED EARTHQUAKE PERFORMANCE

The earthquake performance of a power system can be improved through the adoption and implementation of an earthquake preparedness program. Earthquake preparedness programs can be divided into two types of activities: earthquake response planning and earthquake mitigation. Earthquake response planning activities address issues that improve the response after an earthquake occurs. These activities are directed at creating disaster response plans or making additions to or changes in physical facilities that will aid in postearthquake response. Examples include enhancing communication capabilities, creating an emergency operations center, or creating an alternate energy control center. Earthquake mitigation activities are directed at improving earthquake performance by reducing damage and facilitating repairs after an earthquake.

Although disaster response plans are important, their detailed development is outside the scope of this document. A few comments on disaster response plans are given below. Figure 4.1 shows a flow chart of the relationship of the steps in formulating and implementing an earthquake preparedness program.

4.2 EARTHQUAKE HAZARD AND SYSTEM VULNERABILITY EVALUATION

An initial hazard evaluation is needed to assess the potential for occurrence of damaging earthquakes in the utility's service area. An initial assess-

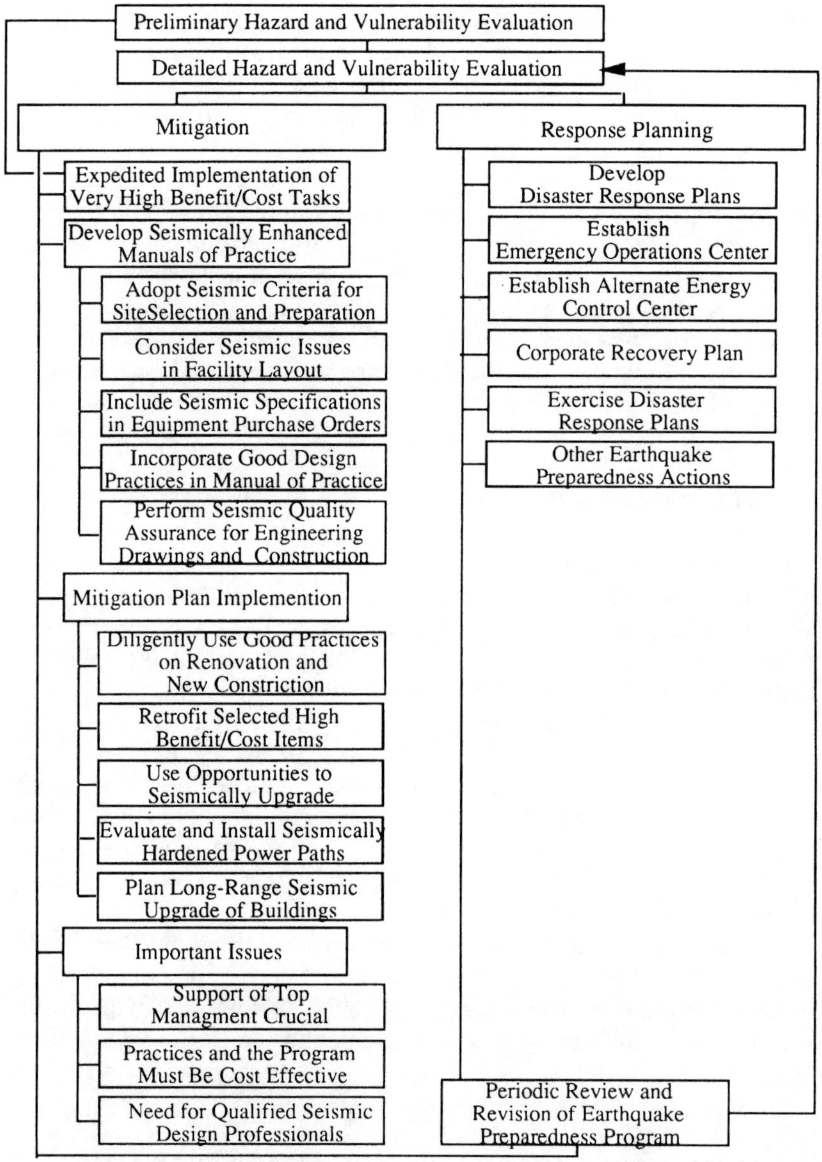

FIGURE 4-1. Flow Chart of the Steps in Formulating and Implementing an Earthquake Preparedness Program.

ment of the vulnerability of power facilities is needed to establish the scope of earthquake risks and the need for more detailed hazard and facility evaluations. In many situations, there will be a need for a detailed hazard and vulnerability evaluation before disaster planning and mitigation can proceed.

4.2.1 Initial Earthquake Hazard and System Vulnerability Evaluation

One of the first steps in determining whether an earthquake preparedness program is necessary is to perform an initial evaluation of the earthquake hazard and power system vulnerability. Should a preparedness program be justified, the initial evaluation is needed to determine the scope and character of the risk and the types of measures that will be needed to carry out an effective preparedness program. The goal of the initial evaluation should be to get rough, realistic, and conservative estimates for direct losses. The procedures for performing an evaluation of earthquake hazards are described in Section 4.2.2. There may be some situations where the cost and effort associated with a detailed evaluation may not be necessary or justified. In such cases, the methods described in Sections 4.2.2.1 and 4.2.2.2 may provide sufficient information to proceed with the response planning and mitigation parts of the program.

Information provided in Chapters 5 through 10 allows system vulnerabilities to be assessed. After the evaluation, the earthquake response planning and mitigation programs can be formulated.

After the initial evaluation of hazards and vulnerabilities, certain mitigation tasks may be obvious because of their high benefit-cost ratio. For these tasks, retrofitting can proceed without formulating an entire mitigation plan. Examples of such tasks include providing adequate and earthquake-secure emergency power at generating stations and substations.

4.2.2 Detailed Earthquake Hazard and System Vulnerability Evaluation

The objective of this effort is to gather data to make a case to management for initiating a formal earthquake preparedness program. For many utilities, the losses associated with a great earthquake that impacts their service area will have very significant financial impacts that responsible management should not ignore. The evaluation includes the following tasks:

- determine whether the earthquake hazard warrants a preparedness program and establish seismic zones in the service area;
- obtain a rough assessment of direct damage and associated costs;
- estimate the extent and duration of power disruption;
- estimate damage reduction to be achieved by a long-term mitigation program;

- identify the benefits to be achieved in the area of risk management; and
- identify ancillary benefits derived from a mitigation program.

Each of the above tasks is discussed in more detail in the following sections.

4.2.2.1 Determine Whether the Earthquake Hazard Warrants an Earthquake Preparedness Program and Establish Seismic Zones in the Service Area. The revised version (1998) of IEEE Standard 693[1] and the adoption by the three major model building codes of the same seismic hazard maps greatly simplifies the process of determining whether a utility is exposed to earthquake hazards. The IEEE Standard 693 references the 1987 NEHRP requirements; however, it actually uses the 1994 NEHRP edition. For consistency with IEEE Standard 693, this document will also use the 1994 NEHRP requirements. However, the seismic hazard maps used with the 1994 NEHRP requirements are outdated and do not consider the risks associated with the subduction zone in the northwest United States.

A comparison of the utility's service area with the hazard maps will indicate the seismic exposure potential. The procedure suggested in IEEE Standard 693 for evaluating a site has been adapted to evaluate a utility's service area. This procedure can also be used to establish seismic zones within the utility's service area. Some California utilities, because of their large service areas, have classified their service areas relative to seismic hazards. Facility design coefficients vary depending on the seismic hazard at the site. Generally, even if the hazards vary within the service area, uniform seismic equipment specifications are used so that equipment can be moved to any site within the service area. A detailed earthquake hazard and system vulnerability evaluation generally includes the following tasks.

- Determine the NEHRP[2] soil type classification that best represents the service area. For large service areas, it may be appropriate to divide the area into different seismic vulnerability regions based on soil types and expected peak ground accelerations given on risk maps. In characterizing the soil classification that best represents the service area or region, the soil conditions at transmission substations are the most critical. Added significance is given to transmission substations that are particularly important to system operations. The five soil types (A, B, C, D and E) defined in the NEHRP document are summarized in Table 4-1. For a complete description, see endnote 2, Section 4.1.2.1.
- Using the NEHRP seismic hazard map with 2% probability of exceedance in 50 years on soft-rock sites, locate facilities in the service area. This map gives the response acceleration (in $\%g$) for short and at 1-second periods.
- Estimate the mapped spectral response acceleration at short periods from the 2% map at sites in the service area.

TABLE 4-1. Soil Profile Type Classification.

Soil Profile Type	Soil Description
A	Hard rock with measured shear wave velocity larger than 1,500 m/s (5,000 ft/sec)
B	Rock with shear velocity above 760 m/s (2,500 ft/sec) and equal or less than 1,500 m/s (5,000 ft/sec)
C	Very dense soil and soft rock with a shear velocity above 360 m/s (1,200 ft/sec) and equal or less than 760 m/s (2,500 ft/sec)
D	Stiff soil with a shear velocity equal or above 180 m/s (600 ft/sec) and equal or below 360 m/s (1,200 ft/sec)
E	A soil profile with a shear velocity less than 180 m/s (600 ft/sec)

Note: Adapted from endnote 2, Section 4.1.2.1, Site Class Definitions.

TABLE 4-2. Values of F_a as a Function of Site Conditions and Shaking Intensity.

SoilProfile Type	Shaking Intensity (A_a)				
	$A_a \leq 0.1$	$A_a = 0.2$	$A_a = 0.3$	$A_a = 0.4$	$A_a \geq 0.5^a$
A	0.8	0.8	0.8	0.8	0.8
B	1.0	1.0	1.0	1.0	1.0
C	1.2	1.2	1.1	1.0	1.0
D	1.6	1.4	1.2	1.1	1.0
E	2.5	1.7	1.2	0.9	—[b]

Note: Adapted from endnote 2, Table 4.1.2.4a. Use straightline interpolation for intermediate values of A_a.

[a] Values of $A_a > 0.4$ are applicable to the provisions for seismically isolated structures.

[b] Site-specific geotechnical investigation and dynamic site response analyses shall be performed.

- Multiply these values by the site coefficients, F_a, defined for the site soil conditions as a function of spectral response accelerations (A_a) at short periods given in Table 4-2.
- From the resulting values, which are the site peak ground accelerations, the performance level in IEEE Standard 693 is determined. If the peak ground acceleration is equal to or less than 0.1*g*, the site is seismically classified as *Low.* If the peak ground acceleration is greater than 0.1*g* but equal to or less than 0.5*g*, the site is classified as *Moderate.* If the

peak ground acceleration is greater than 0.5*g*, the site is classified as *High*.

4.2.2.2 ***Obtain a Rough Assessment of Direct Damage and Associated Costs.*** The most realistic and accurate method of estimating power system damage is to base it on scenario earthquakes. In scenario earthquakes, the locations and magnitudes of hypothetical earthquakes are selected, and system damage is evaluated for each event. However, for the purpose of this initial estimate, it is suggested that the seismic exposure be obtained directly from the risk maps because it is much simpler and reduces the expertise needed for implementation.

Depending on the size and location of the utility's service area, damage estimates using the risk maps may be high. However, utilities with small service areas should obtain good damage estimates. The primary reason that the use of the risk map may overestimate damage is that the peak accelerations shown on the risk map may be derived from several earthquakes, and the accelerations caused by a single earthquake may not impact the entire service area. One method of judging whether damage estimates are high is to compare the utility's service area to isoseismal maps given in Chapter 2. If a significant part of the service area is covered by one of these historical earthquakes or by similar isoseismals obtained by shifting the epicenter slightly, damage estimates should be reasonable, as this indicates that a single event can impact the entire service area. The purpose of this rough damage estimate and the resulting direct cost of damage is to determine whether a cost-effective program should be started. These estimates would not be used to determine specific mitigation actions. Before specific mitigation actions are taken, all vulnerable sites should be evaluated.

Past experience has shown that engineers who design power facilities and who have not previously given much attention to seismic issues may need assistance in assessing the earthquake vulnerability of their facilities. The use of historical data is vital in getting realistic estimates of damage. For the purposes of an evaluation, a careful review of the earthquake performance of power system equipment contained in this document should provide the information needed to identify vulnerable equipment. The initial review should focus on transmission substations, as these have been where most earthquake damage has occurred. This review should reflect seismic practices used by the utility. It is not envisioned that all facilities would be given a walk-through. After the document is reviewed and understood, a few key facilities should be physically reviewed to assess the seismic practices used by the utility. The assessment should take into account the fact that the performance of facilities is based primarily on damage observed within California, where seismic practices have been evolving for many years. Based on estimates of damage at the sites that were visited, damage estimates for the entire system can be made by extrapolation.

Once a list of potential equipment damage has been made, the repair and/or replacement cost of the damage can be estimated by using data from recent purchases or older purchases adjusted for inflation. The cost of damage should include the cost to repair or replace equipment and the cost of installation materials and labor.

Prior to formulating a detailed mitigation plan, the vulnerability of all facilities in the service area should be evaluated. The detail of the review should be based on the importance of the facility to employee safety and potential impact of system operations if it were damaged.

4.2.2.3 ***Estimate the Extent and Duration of Power Disruption.*** Even when accurate estimates of equipment damage are available, there is no simple, accurate method for estimating the extent and duration of service disruption. Because of the high degree of redundancy in substations and networks, it is difficult to estimate the impact of equipment damage. Estimating the duration of disruption is difficult because the effectiveness of emergency operating procedures is unknown. The rough character of the damage estimates and the difficulty in determining their impact on disruption make the usefulness of these estimates questionable. The main purpose for estimating the extent and duration of disruption is to get an idea of the potential impact on the community and the size of indirect losses. For the purpose of justifying the need for a mitigation program, estimates of the extent and duration of disruption are probably not necessary. They are only listed here for completeness. Because of the difficulty in obtaining these estimates and because of the uncertainty of these estimates, it is suggested that no estimates be made.

If an estimate of the extent of disruption is pursued, fragility data for the equipment, if available, should be used to determine the likelihood of operability of the equipment after an earthquake. If fragility data are not available, it should be assumed that the most vulnerable equipment will be inoperable if it is exposed to ground motions above $0.2g$. Substation transformers (for instance, those pedestal-mounted on rails, with poor anchorage) and high-voltage (220 kV and above) live-tank circuit breakers should also be considered vulnerable. The effect of some equipment damage can be quickly eliminated by bypassing it, as discussed elsewhere in this document. This approach should be taken into account when estimating the extent and duration of disruption. To get a conservative estimate of the extent of disruption, all suspect circuit breakers, transformers, and other vulnerable equipment in the impacted zone can be circled on a system operating diagram and assumed to be damaged and nonfunctional. This information can be provided to the chief dispatcher, who should be able to provide a rough estimate of the expected disruption.

It is even more difficult to estimate the duration than the extent of disruption. However, the service manager may be able to make a rough estimate of the duration of the disruption, based on the number and location of damaged items, the number of spares, typical repair times, and emergency operating procedures that are expected to be used. Again, it is important to realize that much of the damaged equipment will probably be bypassed to restore service quickly.

In estimating the extent and duration of disruption based on damage estimates obtained from seismic maps, it is important to take into account that the entire service area would probably not be impacted by a single event. Thus, judgment must be exercised in determining the impacted areas. Historic isoseismal maps given in Chapter 2 are useful in providing guidance. In many parts of the country, utility service areas are much smaller than those of the larger California utilities; hence, many of these utilities may have most or all of their system directly impacted by a single event.

4.2.2.4 Estimate Damage Reduction To Be Achieved by a Long-Term Mitigation Program. Damage reduction results from better seismic practices applied to facilities over a period of several decades, as older facilities are refurbished and new facilities are constructed. Because of the large physical plant characteristic of most utilities, it will take a long time after good practices are instituted before a significant part of the system incorporates good seismic practices. The implementation of IEEE Standard 693 should improve the seismic performance of substation equipment. However, utilities must be diligent in the use of seismic specifications in purchase orders and the application of recommendations contained in this and other guides. Reference should be made to Sections 4.4.4 and 4.5.5.

Evaluation of long-term benefits should take into account facilities that are not normally upgraded, such as substation control houses and other utility structures. The failure of these structures can have an impact on disruption that is significantly greater than the direct losses associated with their damage.

4.2.2.5 Identify the Benefits To Be Achieved in the Area of Risk Management. One of the benefits of improved earthquake resistance of facilities and a realistic assessment of their true vulnerability is the potential reduction in insurance costs associated with risk management. Technical guidance in this area should be available within the risk management group at the utility.

4.2.2.6 Identify Ancillary Benefits Derived from a Mitigation Program. Ancillary benefits that may be derived from improved earthquake resistance include reduced loss of service, fewer injuries to personnel, reduced liability for effects of service disruption, and reduced risk of governmental intervention resulting from a poor system response. These cost savings, over the long term, may be much larger than those from preventing direct losses due to equipment damage in a given earthquake. However, ancillary benefits are difficult to quantify.

4.3 EARTHQUAKE PLANNING

Power companies are organized and have experience to manage small and moderate emergencies that may occur often. On an annual basis, major storms can elicit a major emergency response. Because of this, there may be a tendency to think that earthquakes can be addressed as just another of the many disasters—such as floods, wind storms, or ice storms—visited upon power systems. However, earthquakes impact a utility in a significantly different manner than other disasters. The major difference between earthquake damage and damage due to more common emergencies is that substations, rather than distribution lines, are damaged. The scale of a disastrous earthquake-related emergency will far exceed any of the emergencies that most emergency response plans are designed to handle. Most utility personnel will never experience a major earthquake emergency in their entire professional career with a utility. The damage and disruption caused by a large earthquake, or by a very severe hurricane like Andrew, require a disaster plan that focuses on the unique problems associated with earthquakes and very severe storms. These unique problems include damaged substation equipment and disruption to transportation, communication, and other lifeline systems.

The development of a disaster response plan is particularly important in the early stages of a long-term earthquake mitigation program, because the utility is most vulnerable during this period. Because each type of disaster, be it a large earthquake, a 500-year flood, or a 100-year ice storm, has unique features, some aspects of the plan should address each of these unique issues. The disaster response plan may require the development of some new facilities, such as an emergency operations center.

A good plan will prove invaluable in managing an emergency. An important added benefit in developing emergency response plans is derived through the discussion of issues and the interaction of the players and stakeholders in the generation of the plan. Earthquake planning, however, is more than an emergency response plan. The plan has to be exercised, and support facilities, such as the emergency operations center and enhanced communications, must be developed.

4.3.1 Disaster Response Plans

Emergency response plans should be reviewed from the perspective of the unique problems associated with earthquakes. Because of the potential magnitude of the damage and disruption following a major earthquake, response plans are referred to here as disaster response plans to distinguish them from the other emergency response plans that most utilities already have in place. The term "disaster" implies a much larger and more damaging event. While disasters are not limited to earthquakes, as demonstrated by hurricane Andrew, the issues that should be addressed in a disaster response plan far exceed those addressed by even severe emergencies.

Many items in a disaster response plan will be the same for different types of disasters. The need to plan and develop an emergency operations center, the need to enhance communication systems, and the need for a corporate recovery plan will be similar for most disasters.

Disaster plans should be exercised annually. One value of such exercises is that flaws in the plan may be identified. Another is that individuals who participate in the exercise will be reminded of what roles they will play in a disaster response situation and will have an opportunity to interact with and get to know other participants, whom they normally would not encounter. Finally, management can be brought into the exercise so their awareness and support can be maintained.

4.3.2 Corporate Recovery Plans

A major earthquake may have a large impact on business operations as well as on electrical operations. For example, large, unanticipated capital expenditures may be required to repair damage to the system, and system or customer damage may prevent power deliveries and cause a loss of revenue. Recent earthquakes provide examples of actual direct losses to power utilities to repair or replace damaged equipment and restore service. The 1989 Loma Prieta earthquake, centered south of San Francisco, caused about $100 million in damage, and the 1994 Northridge earthquake in southern California had over $180 million in damage. Not included in these figures are other financial losses due to the loss of revenue from system disruption and the need to purchase expensive make-up power. Clearly, corporate finance is directly and significantly affected, and adequate planning for the impacts of a major earthquake can improve the utility's response to such events.

4.3.3 Evaluation of System Vulnerabilities

Most mitigation measures are directed at details, such as equipment anchorage and installation practices. There is also a need to look at broader system issues, such as the vulnerability of particularly important substations, the energy control center, and the number and storage of spare parts.

4.3.4 Emergency Operations Center

The disaster response plan will require the creation of an emergency operations center. The experiences of the utilities after hurricane Andrew and following the Loma Prieta and Northridge earthquakes have advanced the state-of-the-art in dealing with disasters, and many lessons have been learned. As a result, changes have been made in the organization and operation of emergency operations centers, staffing, and physical layout.

4.3.5 Alternate Energy Control Center

California utilities have created alternate energy control centers in seismically less vulnerable locations to enhance system reliability. If a building housing an energy control center appears to have structural damage, building inspectors may "tag" the building as unsafe until a detailed inspection is performed. Notwithstanding the pleading of utility management, access to critical facilities almost certainly will be denied by police. Of course, if a structure is severely damaged, a utility would not consider putting their personnel at risk.

Some energy control centers would be difficult for personnel to access if the earthquake damages the transportation system, as might be expected. If transportation is disrupted, an alternate energy control center could be used even if the primary center is not damaged. The selection of an alternate control center should consider ready access to the facility by utility personnel after an earthquake.

4.4 EARTHQUAKE MITIGATION

The implementation of an earthquake mitigation program will lead to a progressive improvement in the seismic ruggedness of power facilities in a cost-effective and manageable way and will improve the earthquake response by reducing damage. The approach suggested in this document is to initiate the use of good seismic design practices for new facilities and older facilities undergoing normal renovation. Simple measures can be used that have been shown to be effective in improving seismic performance. For certain facilities or equipment, for which malfunction would be very disruptive, expedited seismic retrofit may be justified. Finally, there are some power system elements that are not routinely refurbished, such as the structural systems in substation control houses and the building that contains the energy control center. The seismic risks of these facilities should be evaluated, and a long-term seismic upgrading program established.

Several of the tasks that make up a mitigation program are generally carried out in parallel. Examples include updating the manual of practice, identifying system vulnerabilities, retrofitting items with high benefit-cost ratios, and incorporating good seismic practices in normal renovation and new construction.

Because retrofitting facilities is, in general, not very cost-effective, the number of retrofitting tasks will probably be relatively small. One common, cost-effective, retrofitting task is the anchoring of transmission transformers.

The majority of the effort to improve the earthquake performance of the system will be done in the course of normal system renovation. The key to this process is to establish and implement good earthquake practices for new facilities and for routine upgrading of older facilities.

The elements of a mitigation program to reduce damage are the following:

- upgrading manuals of practice to incorporate good seismic design practices;
- assessing the vulnerability of power system facilities;
- implementing a mitigation plan; and
- periodically reviewing and revising the mitigation program.

The seismic mitigation program involves many tasks, some of which may be performed in parallel. For a given task—for example, retrofitting items with a high benefit-cost ratio—one could agonize over the optimum order in which different items should be retrofitted. An issue more important than prioritization is that a program be started. Prioritization of mitigation actions will be significantly influenced by existing design and installation practices and by the perceptions of the individuals managing and implementing the program. Upgrading the manual of practice should be given high priority so that good installation procedures are used for new construction.

4.4.1 Implementing Tasks with a High Benefit-Cost Ratio

After the evaluation of hazards and vulnerabilities, certain mitigation tasks may be obvious because of their high benefit-cost ratio; for these tasks retrofitting can proceed without formulating an entire mitigation plan. Some vulnerabilities exist for which the benefit-cost ratio is so large that a detailed analysis of risk is not necessary. An example is provision of adequate backup power at power-generating stations. Emergency batteries or standby power at a generating station play a vital role immediately after an earthquake. It is very common to lose off-site power, and damage to switchyards may force the generating unit off line. If this occurs, power will not be available to operate the bearing-lubricating pumps and the turning gear as the turbine slows. Without adequate lubrication, the main bearings will be damaged. The loss of the turning gear will result in deformation of the turbine shaft as it cools. These unfortunate events have occurred at more than one facility following moderate and small earthquakes. A simple review of the system may identify these items that should be given very high priority for correction.

4.4.2 Seismically Upgrading Manuals of Practice

The manual of practice as defined in this Manual is the collection of a utility's standards to be used for site evaluation, selection, and preparation;

standards for substation layout; standard clauses incorporated into equipment purchase order specifications; detailed engineering drawings of equipment anchorage; and conductor-stringing practices in substations.

It is vital that good seismic design practices be explicitly defined and incorporated into the manual of practice. Good design details for the installation of equipment in manuals of practice set the stage for long-term improvements in the earthquake response of a utility. It is the institutionalization of good practice into the normal practice of a utility that will assure the long-term success of the earthquake mitigation program.

4.4.2.1 Site Selection and Preparation. Many factors must be considered in the selection of a power facility site. In the past, seismic vulnerability has generally not been a high priority or even a consideration. However, it is important that seismic vulnerability be considered in power facility site selection. In many areas of the country, the future expansion of power facilities will be at existing sites. It is important that seismic vulnerability be given higher priority than currently at these sites as they are expanded and become more important to the utility's earthquake performance. If the cost of site remediation needed to make the site seismically acceptable is incorporated into the selection process, other less vulnerable sites may become preferable.

On sites that require cut-and-fill in their preparation, proper compaction on fill parts of the site must be performed, and discontinuity at the boundary between the cut and fill should be considered in arranging the buildings and equipment on the site.

Consideration of these aspects of site selection and design should be incorporated into the manual of practice.

4.4.2.2 Facility Layout. The manual of practice should identify features in the facility's layout that may impact its earthquake performance. For example, the potential for soil deformation at boundaries between cut-and-fill areas and due to lateral spreading should be incorporated into facility layout. Other features to consider in the site layout include the spacing of equipment so that adequate bus flexibility can be provided and excessively long conductor runs can be avoided.

Manuals of practice should also mandate that seismic review of site layouts be done early in the design cycle so that good seismic practices can be incorporated before planning is completed.

4.4.2.3 Seismic Specifications for Equipment. The establishment and implementation of IEEE Standard 693 greatly simplifies the process of setting seismic specifications for substation equipment and of obtaining equipment that satisfies the specifications. This standard provides guidance for the seismic qualification of substation equipment.

Utilities should provide seismic specifications or select equipment that is known to have good seismic performance for new or upgraded facilities. For example, key circuit breakers that form a critical power path through the facility should be given special consideration for important facilities. IEEE Standard 693-1997 provides a simpler process than was historically used, where qualification was done on a site-by-site basis.

*4.4.2.4 **Installation Practices.*** Many of the changes to improve the manuals of practice will be related to methods of installing equipment. Manuals of practice should explicitly specify details that are important to earthquake performance of facilities. Installation practices include proper torquing of bolts, leveling equipment, and conductoring. Anchorage drawings for equipment should be based on the Seismic Outline Drawings required in IEEE Standard 693 that specify anchorage loads. Drawings should specify the details of how conductors are to be connected to the equipment, including provisions for connection flexibility. Examples of good seismic design practices are provided in this Manual.

*4.4.2.5 **Seismic Quality Assurance.*** Individuals as well as organizations are resistant to change: they tend to do things in the future as they have done them in the past. Improved earthquake performance will require that new practices be used to implement the lessons learned from earthquake damage. Thus, seismic quality assurance must be done by specialists to check engineering drawings and field construction to assure that current, well-thought-out seismic practices are being followed.

At the engineering level, engineering drawings should be given a seismic review. That is, an engineer who is familiar with good seismic practices should review the engineering drawings to assure that they reflect current practices related to site layout and equipment installation. In addition to incorporating the recommendations put forward in this document, it is important to review the load path that is needed to resist seismic-induced loads.

It is particularly important to inspect construction for conformance to new seismic practices during the initial phases of an earthquake mitigation program to assure that the new practices are actually used. For example, tradespeople may use older "standard" installation details without referring to new drawings; unless an effort is made to monitor new practices, older, inadequate installation practices may continue to be used.

Engineering seismic specialists (see Section 4.5.5) should review the seismic qualification reports carefully to assure that the equipment meets the seismic specifications. Historically, many utilities have assumed that the reports were correct and adequate. The reports should be checked by a seismic specialist within the utility or by an independent representative designated by the utility.

4.4.3 Detailed Vulnerability Assessment of System Facilities

Before a mitigation plan can be implemented, a detailed assessment of the facilities at risk must be done. This will require that all facilities be evaluated. However, not all facilities need be evaluated prior to initiating a mitigation plan. Many facilities have historically performed well, and their evaluation can be postponed to later in the implementation process. Examples of facilities for which the evaluation could be delayed might include transmission towers, low-voltage substations, and power-generating stations. A selective assessment of these low-risk facilities may be appropriate. For example, a review of the anchorage of power transformers at large low-voltage substations and at power-generating stations may be appropriate.

Since the proposed approach calls for relatively little retrofitting, only those vulnerable items that might require retrofitting need to be evaluated. For example, backup power and the anchorage of power transformers at substations and generating stations should be evaluated. Thus, not all equipment in the entire system needs to be evaluated before a mitigation implementation plan is formulated.

4.4.4 Implementation of Mitigation Plan

Several mitigation tasks can be identified. Some will be executed in parallel, and others will be implemented later in the program, as funds become available. For example, manuals of practice may be upgraded in parallel with disaster planning, while good practices are used as new facilities are constructed in the future.

4.4.4.1 Incorporate Good Seismic Practices in Normal Facility Renovation and New Construction. It is vital that the renovation of existing facilities and new construction incorporate good seismic practices. This is the fundamental premise for the suggested approach. Thus, it is important that manuals of practice be updated to reflect good practice.

4.4.4.2 Retrofit of Selected High Benefit-Cost Items. Items with a high benefit-cost ratio should be retrofitted. Examples are the anchoring of emergency batteries and transmission transformers. Because of the low cost, retrofitting inadequate anchorage is often cost-effective. However, the decision to retrofit equipment should be made on a case-by-case basis where the cost of retrofitting and consequences of failure are considered.

4.4.4.3 Considering Seismic Vulnerability in Prioritizing Normal Renovation Schedule. High priority should be given to replacing equipment that is known to perform poorly and the failure of which would cause secondary damage or large disruptions.

4.4.4.4 Taking Advantage of Upgrading Opportunities That Present Themselves. Make minor upgrades that can improve seismic performance in the course of normal maintenance as opportunities present themselves. For example, the performance of some friction-clip equipment anchors has been poor. Welding of the clip to the equipment can significantly improve its performance. Thus, when welding equipment is scheduled to be used at a specific substation, this opportunity should be used to weld the friction clips on the most vulnerable equipment.

4.4.4.5 System Review and Seismic Hardened Power Paths. Some California utilities have implemented selective upgrading of some of their facilities to improve system earthquake performance. An evaluation of the system may indicate that one or more critical substations are vulnerable to damage. Rather than replace the entire substation, the seismic hardening of a critical power path through important substations can provide significant protection at relatively little cost. This will reduce the risk of long-term disruptions and may even avoid disruptions entirely. For example, at a key 230-kV to 115-kV substation, the vulnerable 230-kV circuit breakers and disconnect switches that feed the 115/230 transformer bank could be replaced with seismically robust units. As a result, the 115-kV bus could remain energized. As will be discussed latter, the 115-kV system is inherently seismically rugged.

4.4.4.6 Upgrading Facilities Not Normally Renovated. In the normal course of development, some facilities are not renovated. For example, the structural systems of control house buildings and the structure that houses the energy control center are usually not modified when the equipment they contain is periodically replaced. These and other structures that affect personnel safety should be upgraded if necessary. Examples include service center buildings, corporate offices, and structures that house the engineering departments and the trouble board operations. It is necessary to identify risks associated with structural vulnerabilities and establish a long-term seismic upgrading program.

4.4.5 Periodic Review and Revision of Mitigation Program

Any program that is designed to last for several decades will need to be periodically reviewed and revised. Earthquakes may provide new lessons that will need to be incorporated into manuals of practice. The improved understanding of earthquake risks may change the urgency with which mitigation measures should be implemented, as recently occurred in the Northwest with the discovery that large subduction earthquakes have occurred and that there is the potential for a Magnitude 9 earthquake. It is suggested that the program be formally reviewed every five years unless changing circumstances require more frequent review. This review should determine if there has been any significant change in the understanding of the seismic hazards, if recommended seismic practices have changed, and if, because of the evolution of equipment design, there are differences in the seismic ruggedness of equipment that would change the vulnerability of the power system.

4.5 COMMENTS ON IMPLEMENTING AN EARTHQUAKE DAMAGE MITIGATION PROGRAM

4.5.1 Initiating an Earthquake Mitigation Program

Contemplating the initiation of an earthquake mitigation program for a utility can be overwhelming. Most utilities have a physical plant with a replacement value of billions of dollars, most of which was constructed with little or no consideration given to earthquakes. Because of limited resources, it is very important that the mitigation program be approached in a gradual and orderly manner so that meaningful inroads can be made toward improving earthquake performance. It should be emphasized that the approach suggested here may not be the optimal or most efficient way to address the problem for every situation. The resources needed to conduct an evaluation of the system to optimize an earthquake mitigation effort are substantial and may not be consistent with the low-cost methods suggested here. What is important is that the effort be started and sustained.

4.5.2 Commitment of Top Management Is Needed

While decisions to improve the earthquake resistance of power systems can be made at lower administrative levels within an organization, the support of top management is required to institutionalize a mitigation program. Changes in design practices and equipment specifications, development and exercise of disaster response plans, and changes in manuals of practice all require high-level management decisions. The best approach is to get top management involved and committed from the start of the effort. Getting management support requires that the vulnerabilities, mitigation measures, and the cost as well as benefit of a mitigation program be understood and communicated to management in terms that are understandable and relevant to its perspective.

In approaching management, it is important to emphasize the differences in the proposed methods for dealing with earthquake risk mitigation and those associated with the nuclear power industry. Unfortunately, measures to reduce the impact of earthquakes can carry with them a history of seismic problems encountered in nuclear power plant construction. In many cases, seismic requirements associated with nuclear plants were costly and, in some cases, even counterproductive. Also, potential difficulties in dealing with outside organizations, such as the Nuclear Regulatory Commission, are avoided. It is vital to emphasize that the proposed earthquake mitigation program is under the control and direction of the utility and is based on cost-effectiveness. The problems that many utilities had relative to earthquake requirements in construction of nuclear generating facilities are not applicable to the facilities of concern here.

To justify an earthquake mitigation program, detailed benefit-cost analysis associated with the cost of repairing or replacing equipment, the potential lost revenues, and the indirect cost to the community could be performed. However, detailed analysis of all elements is not necessary prior to initiating a program. First, the cost of doing detailed, accurate benefit-cost studies may be comparable to the initial stages of implementing the program. In many regions, the probability of scenario events that are considered is so low that costly efforts cannot be justified.

Many of the mitigation measures suggested here are low cost and can be justified on the basis of "engineering judgment," rather than formal analysis. It is for this reason that the emphasis is on new construction rather than retrofitting. However, there are many cases where retrofitting can be justified. The loss of station batteries at a generating station can result in damage to turbine bearings and turbine shaft distortion. A detailed analysis is not required to compare the cost of a seismically adequate battery rack to the damage and downtime of a large turbine costing about $100 million.

4.5.3 Cost-Effectiveness

Two key elements for an effective earthquake hazard mitigation program are as follows: first, the entire program must be cost-effective and, second, the mitigation measures must be institutionalized. In this way, good practices become part of standard design and construction practices so that a crusade is not required for every mitigation measure. While a continuing effort will be needed to update methods and address issues as new information becomes available, much of the program should be self-sustaining once it is instituted. For example, if a policy is established to anchor transformers for new construction, it would over time yield a system in which all transformers are anchored.

The key to a cost-effective mitigation program is a strong emphasis on improving the earthquake resistance of new facilities and incorporating good practices in routine upgrades and refurbishment. Retrofit activities should be limited to those items that pose an unacceptable risk in light of their high benefit-cost ratio.

4.5.4 Maintaining Mitigation Program

In recent years, California has had the "advantage" of experiencing a damaging earthquake every few years. Thus, earthquake issues continually reassert themselves. In regions with less seismic activity, the long-term maintenance of a mitigation program is more difficult. The support of utility management is vital to the long-term maintenance of an earthquake mitigation program. In the early phases of the program it is important that management and other groups, such as procurement, be informed of the importance and benefits of earthquake mitigation.

Three things can be done to maintain the momentum of a mitigation program.

First, the program should be institutionalized by ensuring that critical features are embedded in manuals of practice. In this way, the characteristic of organizations to resist change works to the advantage of the mitigation program.

Second, it is important that oversight of the program be a major responsibility of one or more individuals, so that there is someone to continue to pursue the program objectives. In less seismically active areas and in smaller utilities, oversight of an earthquake mitigation program may not be the sole responsibility of a single individual. However, it is vital that the individual(s) have management support and that all employees, office and field personnel alike, know that the seismic specialist's recommendations should be followed. In the current utility environment of downsizing and increased workloads, more immediate problems may push aside the earthquake mitigation program.

Third, the mitigation program should have well-identified milestones with a schedule for their completion.

An important part of the maintenance, effectiveness, and quality assurance of an earthquake mitigation program would be to train maintenance personnel on issues that relate to good earthquake performance.

4.5.5 Seismic Design Engineering

Seismic design is usually done by a licensed civil engineer (in some states with a specialty in structural engineering, design is done by a licensed structural engineer). Although practices vary among utilities, within the power industry many design features that are important for good earthquake performance of substations are designed by individuals with little understanding of earthquakes or their effects. Worse, certain important tasks are handed off to tradespeople without guidance. For example, the slack and flexibility of conductors connecting substation equipment is often left to the discretion of tradespeople. The design of conductor connections and the selection of porcelain supports is often done by electrical engineers with little consideration given to potential earthquake dynamic response or loads.

It is difficult to specify appropriate credentials for the individual(s) who should be responsible for seismic design. In this document, it is assumed that the design of power facilities or the installation of equipment is to be done by an individual or individuals who are familiar with good seismic design practices used in electric power transmission systems and who take into account issues that affect the earthquake performance of equipment and facilities. This may require several disciplines working together to ensure that the reliability of the system and equipment are evaluated thoroughly.

After a damaging earthquake, damaged equipment is removed as quickly as possible so service can be restored. The in-house availability of knowledgable engineers to assess damage while the evidence is still available is vital so that the industry can learn from earthquake experience.

ENDNOTES

1. Institute of Electrical and Electronic Engineers (IEEE). 1997. *Recommended Practices for Seismic Design of Substations,*. IEEE Standard 693.
2. Federal Emergency Management Agency. 1995. *1994 Edition: NEHRP Recommended Provisions for Seismic Regulations for New Buildings, Part 1: Provisions.*

Chapter 5
SUBSTATIONS

Recommendations: Retrofitting and New Construction

Seismic impact loads due to flexible equipment anchorage, oversize bolt holes, or inadequate flexibility in conductor connections should be avoided (5.5.1).

Equipment anchorage should have adequate strength to resist pull-out and shear loads and to be stiff. The design of an anchor should consider the entire load path from the equipment to the foundations. Anchorage design should consider shear, vertical accelerations, overturning moments, prying action, and torsion due to eccentricities. High-strength bolts should be avoided as ductile action in the bolts is desirable. (5.5.2)

If bolted anchors are used for heavy switchyard equipment, oversize bolt holes should be avoided or welded plate washers should be used, to provide lateral restraint to connections (5.5.2).

The anchorage of control cabinets should be through the frame of the cabinet. Cabinets lacking frames should use plate washers to prevent sheet metal from tearing. (5.5.2)

The placement of anchors and the equipment configuration should avoid prying action on bolts or welds (5.5.2).

Expansion anchors used to anchor control-house equipment racks should be at least 0.5 in. in diameter and satisfy the building code (5.5.2).

The use of cast-aluminum hardware should be designed to take brittle failure into account (5.5.3, 5.6.3.1).

Adequate flexibility should be provided to conductor connections to accommodate earthquake-induced motion of conductor anchor points (5.6.4).

Positioning the upper attachment points on vertical conductor drops can maintain adequate phase-to-phase separation even when generous slack is provided to conductors (5.6.4).

Where site conditions indicate the potential for soil-structure interaction, cost-effective measures should be evaluated to mitigate damage to transformers (5.7.2.1.2).

Power Transformers: Anchorage

- The foundation slab must be thick enough for embedments to develop the needed pull-out strength (5.7.2.5.1).
- The area of the foundation slab or the use of piles must be such that the bearing capacity of the soil is not exceeded. Where appropriate, the soil under the transformer should be compacted prior to pouring the foundation, and poor soils may need to be replaced with engineered fill (5.7.2.5.1).
- The foundation slab dimensions should be designed to withstand overturning moments (5.7.2.5.1).
- The foundation slab must be strong enough to withstand bending moments for that part of the slab that extends outside the transformer footprint (5.7.2.5.1).
- Where conditions for soil-structure interaction exist, various options to reduce potential impacts should be considered. These include providing generous slack to conductor connections, using seismically robust bushings, or maintaining a stock of spare bushings. Under special circumstances, using piers, piles, or other means to control foundation movements can be considered. (5.7.2.1.2)
- Anchorage design should consider the strength and stiffness of the entire load path (5.7.2.2.2).
- Embedment plates must be stiff enough to prevent welds from tearing the plate, to distribute load to headed studs or other means of securing the plate to the foundation slab, and to avoid unacceptable thermal deformation during concrete curing (5.7.2.5.2).
- Voids in concrete trapped under embedments should be avoided (5.7.2.5.2).
- When used for anchoring, studs should be headed rather than "L" or "J" bolts (5.7.2.5.2).
- Welds to embedments should be designed to avoid stress concentrations that could cause the weld to "unzip" (5.7.2.2.3).
- Welds should be designed to accommodate differences in stiffness of the members along the weld length, and bending along the axis of the weld should be avoided (5.7.2.2.3).

- Caution should be used in welding to the transformer case to achieve good penetration without overheating (5.7.2.2.3).
- When anchoring to an existing foundation slab, grouted, adhesive, or undercut expansion bolts should be used (5.7.2.2.3).

Power Transformers: Bushings

- Providing flexible conductor connections to transformer bushings may improve their performance (5.7.3.2).
- For critical circuits, seismically robust bushings should be considered (5.7.3).
- Conductor connections between bushings and lightning arresters should be made so that lightning arrester failure does not adversely affect bushings (see Section 5.9).

Power Transformers: Radiators

- Radiators should be braced to reduce moments on pipe connections that penetrate the transformer case (5.7.4.2).
- For self-supported radiators, connections between the radiator and transformer should have adequate flexibility (5.7.4.2).

Power Transformers: Conservators

- The load path of conservators should be capable of supporting the conservator (5.7.5.3).
- Piping connections between the conservator and the transformer case should have adequate flexibility to accommodate the dynamic motion of the conservator (5.7.5.3).

Power Transformers: Tertiary Bushings and Surge Arresters

- The amount of slack or flexibility of tertiary bus interconnections should be compatible with the deflections of the tertiary bus support structure and rocking of the transformer from soil-structure interaction (5.7.6.2).

Power Transformers: Transfer Busses

- Transfer bus porcelain supports should be able to accommodate the loads associated with the deformation of tall bus support structures (5.7.7.2).

Distribution transformers mounted on pole-supported platforms should be restrained to their platform (5.8.2).

Lightning arrester conductors should be positioned to reduce the risk to transformer bushings should the arrester fail (5.9.4).

Post used to support lightning arresters, conductor, and instrumentation transformers should be provided with stiff base details (5.9.4).

Station transformers should be anchored (5.17.4).

Substation control houses should be included in mitigation plans so that they eventually are upgraded or replaced to conform to modern building code requirements. Special consideration should be given to their postearthquake operability. (5.18.4)

Oil storage tanks should be anchored (5.19.2).

Recommendations: Retrofitting

The use of friction clips for equipment anchorage is not recommended. Welding of the clip to the equipment which it is anchoring can improve its seismic performance. (5.5.2.1)

The use of restraint bands around the porcelain-flange joint of transformer bushings may prevent porcelain from slipping and gaskets from being displaced (5.7.3.2).

The use of transformer lifting lugs should be considered as anchoring points when retrofitting transformer anchorages (5.7.2.2.2).

Recommendations: New Construction

The preferred method of anchoring heavy switchyard equipment is to weld it to foundation embedments (5.5.2, 5.7.2.5).

It is recommended that disconnect switches be tested at the desired performance level for their seismic qualification (5.13.4).

Indeterminate structures fabricated from post insulators, such as large capacitor racks, should use post insulators of the same length or shims to equalize their length (5.16.3).

Transmission and distribution substations have similar types of equipment that serve similar functions, so they will be discussed as a group. In addition, switchyards associated with power-generating stations and switching substations have equipment similar to typical transmission substations and are also included in this section, even though some of these facilities may not have all of the components discussed in this section. For example, switching substations do not have power transformers.

This chapter is organized with introductory sections that provide an overview of the configuration and function of substations and their components, a review of the effects of earthquakes on substations, a brief discussion of design criteria referenced in other standards, and a review of common failure modes of substation components.

There is a section for each of the major classes of equipment. The facilities and equipment classes discussed are busses, conductors and their supports, power transformers, lightning arresters, current transformers, current-voltage and potential transformers, circuit breakers, disconnect switches, circuit switchers, wave traps, voltage support and power factor correction devices, station power, substation control houses and their contents, and miscellaneous facilities. The equipment sections generally have five parts. The first part contains a brief general description of the equipment and its function. The second part reviews earthquake performance of the equipment. The third part discusses mitigation and retrofit methods. The fourth part describes postearthquake emergency operating procedures associated with the equipment. The fifth and last part recommends good practices for the installation of new equipment.

5.1 OVERVIEW OF SUBSTATIONS

Substations serve several functions. They provide protection to transmission and distribution lines and the equipment within the substation. This is done with protective devices that sense abnormal system operating conditions and trigger other devices to isolate the lines or the equipment. Most substations provide for the transfer of power between different voltage levels through the use of power transformers. Substations also provide a means of reconfiguring the power network. There are other functions performed at substations that support the primary operations or meet other power system needs. For example, there are control consoles, station batteries, and monitoring, recording, and communications equipment.

The overall earthquake performance of substations and their equipment has been mixed. Equipment in substations operating at or below 115 kV has performed well when properly anchored. The performance of equipment operating at or above 220 kV is dependent on the specific equipment at substations and installation practices. The performance of equipment operating at voltages between these values is not known, as little or no equipment in this range has experienced significant earthquake excitations.

There have been ten earthquakes since 1971 that have damaged U.S. substations. They have demonstrated the vulnerability of a broad range of substation equipment and facilities. Most of the damaging earthquakes were light or moderate, and damage was limited so that the high level of redundancy designed into substations enabled the substation to continue operation with little or no disruption. There have been two large earthquakes in which several substations were damaged. In a few cases, where substation equipment damage was so great that substation redundancy was overwhelmed, redundancy in the network was able to compensate for the loss of substation functions. There have been cases where there was severe damage to parts of the substation and, to meet network needs, the damaged high-voltage switchyard was bypassed so that transformers could be energized and power restored to the low-voltage switchyard. The measures used by utilities to overcome substation damage have been very effective, and they are discussed below. System performance has been good in the face of significant substation damage because of system redundancies and the ingenuity of utility personnel to work around substation damage.

Soil liquefaction has contributed to substation equipment damage. Most power facility sites that have experienced earthquakes in the United States have had a low soil liquefaction potential. At those sites where liquefaction has occurred, it has not been severe. The potential impact of liquefaction in other parts of the United States, where soil liquefaction potential is much higher, is not known. Soil liquefaction may lower vibration levels; however, subsidence and lateral spreading may impose large interaction loads on equipment conductor connections.

Soil liquefaction should be considered in site selection along with the many other constraints that must be considered. The availability of land that is near power line rights-of-way and other constraints may require that less than optimal sites be used. If the cost of soil remediation is considered in site selection, sites with better soil conditions may be selected. Fortunately, for most substation equipment, foundation loads are low and the lowering of vibration frequencies that accompany soil liquefaction may reduce the effective excitation level experienced by equipment.

The above overall evaluation of substation seismic performance is based only on moderate and large earthquakes. The United States has not yet experienced a major or great earthquake centered in a modern, major metropolitan area.

5.2 SUBSTATION CONFIGURATION AND COMPONENTS

Most substations contain one or more banks of transformers located between high- and lower-voltage switchyards. Switchyards are usually configured in one of four forms: double-bus–breaker-and-a-half, double-bus–double-breaker, double-bus–single-breaker, or ring bus. (These bus configurations are discussed in more detail in Appendix C.) Figure 5-1 shows a schematic diagram of part of a transmission substation that utilizes a breaker-and-a-half bus configuration. All of these designs provide some degree of redundancy, so that if any component fails, service can be quickly restored. The high degree of redundancy has been included to enhance system reliability in the face of many possible contingencies. Similar configurations are used in all parts of the country, so earthquake risk has not been a driving

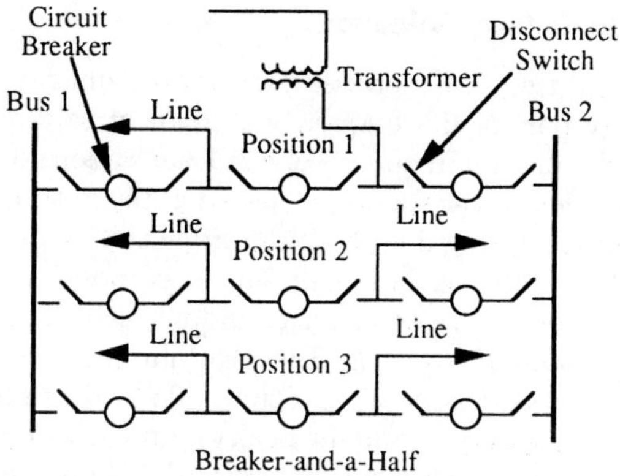

FIGURE 5-1. Schematic Diagram of Part of a Transmission Substation Configured with a Breaker-and-a-Half Bus.

force for this conservative approach, but this design has improved substation seismic response.

Switchyards usually have dead-end structures supporting transmission lines, busses and bus support structures, potential transformers, current-voltage transformers, wave traps, disconnect switches, circuit breakers, current transformers, bus-partitioning devices (circuit breakers or disconnect switches), and a control house. Some systems are configured with two busses connected to one or more transformer banks for enhanced reliability. A transformer bank made up of single-phase transformers often has a spare transformer and a transfer bus so that the spare transformer can quickly replace any of the other transformers.

The physical layout of a substation can influence its earthquake performance. Spacing between equipment can influence the conductor weight that bushings and post insulators have to support. In turn, this can influence conductor design and dynamic interaction between equipment. The location where power is connected to a bus can influence requirements for bus current-carrying capacity. It is important that seismic issues be considered in the early stages of substation layout and design. There have been cases where substation equipment spacing was so close that it was impossible to incorporate adequate flexibility in conductor interconnecting equipment.

5.3 EARTHQUAKE EFFECTS ON SUBSTATIONS

This section will consider the effects of earthquake-induced vibration, soil deformation and ground faulting, and soil-structure interaction.

5.3.1 Earthquake-Induced Vibration

Earthquake-induced vibration has been most damaging to substations. The high-energy part of the frequency content of earthquake vibrations coincides with the natural frequency of much substation equipment. Figure 5-2 shows a free-field accelerogram recorded at the Sylmar Converter Station, Valve Group 7, during the 1994 Northridge earthquake and the resulting ground velocity and displacement. The Northridge earthquake, with a Moment Magnitude of 6.7, was a large earthquake with a duration of severe shaking lasting about 10 seconds. The peak ground acceleration for the ground motion recorded at the substation was 0.93*g*, the peak ground velocity was 117 cm/sec (46 in./sec), and the peak ground displacement was 40 cm (15.7 in.). The response spectrum for this record is shown in Figure 5-3. The response spectra for fractions of critical damping of 0.005, 0.01, 0.02, 0.05 and 0.10 are plotted on a tripartite graph. While many factors can influence the distribution of energy of an earthquake, much of the energy was concentrated in the frequency range from 0.5 to 3 Hz. An introductory text explaining response spectra and earthquake dynamic response of structures is given in endnote 1.

5.3.2 Soil Deformation and Ground Faulting

Soil instability can affect a substation site in several ways. Landslides and rock falls can occur on steep slopes adjacent to a substation, but damage from these sources has not been observed. Subsidence and soil liquefaction can cause the surface of the site to settle, and lateral spreading can cause horizontal movements and cracks in the ground to open. These effects have been observed. Uniform subsidence would probably have little impact on a substation. However, on sites with poor soil conditions, large structures such as bus support structures are often supported on piles or piers that extend to competent strata. In an earthquake, these structures will not settle while slab-supported equipment will settle with the ground. Figure 5-4 shows a low-voltage, gas-insulated circuit breaker with damaged bushings in the Kobe, Japan, earthquake. The conductor drops to the bushings pulled the bushings from the circuit breaker when the ground settled about 30 cm (1 ft). The bus support structure was on piles and did not move. Preventing this type of damage can be difficult, because providing adequate slack to vertical conductors would allow so much lateral movement that phase-to-phase separation may be violated.

Granular, water-saturated, low-cohesive soils can liquefy so that the soil loses its bearing capacity. Lateral spreading can cause large movements that open up cracks in the soil surface. Figure 5-5 shows a lightning arrester and its support structure tilted because of deformation of the soil. A more striking example of foundation movements when soil has lique-

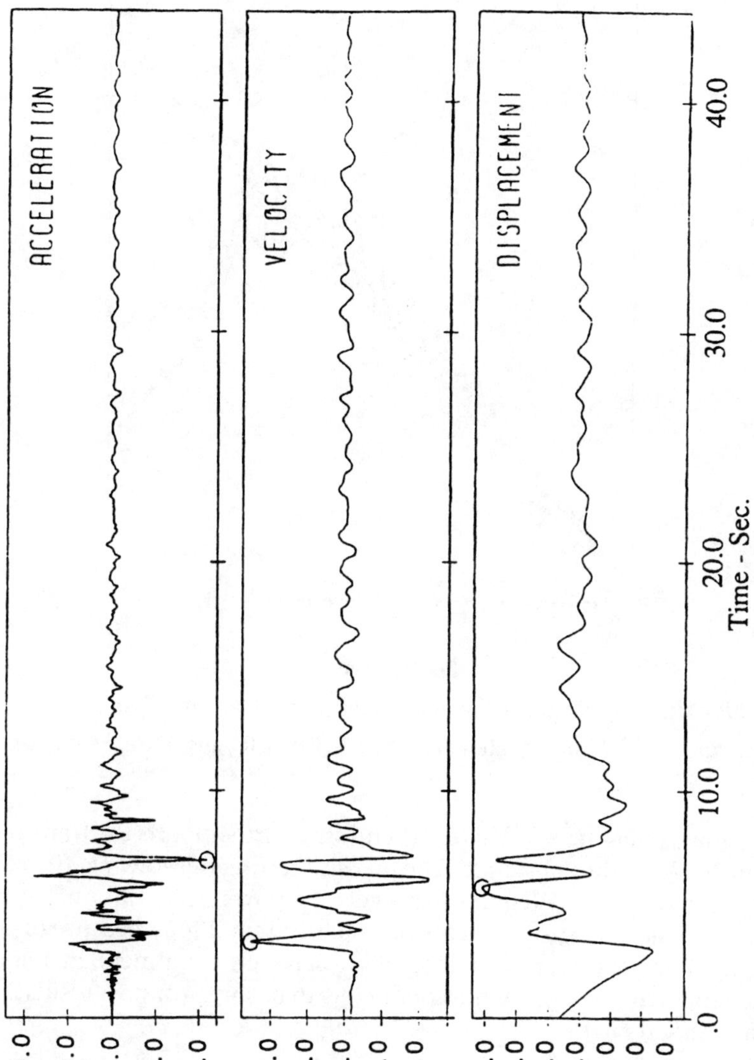

FIGURE 5-2. Acceleration, Velocity, and Displacement Time History of Ground Motion Recorded at the Sylmar Converter Station during the Northridge Earthquake.

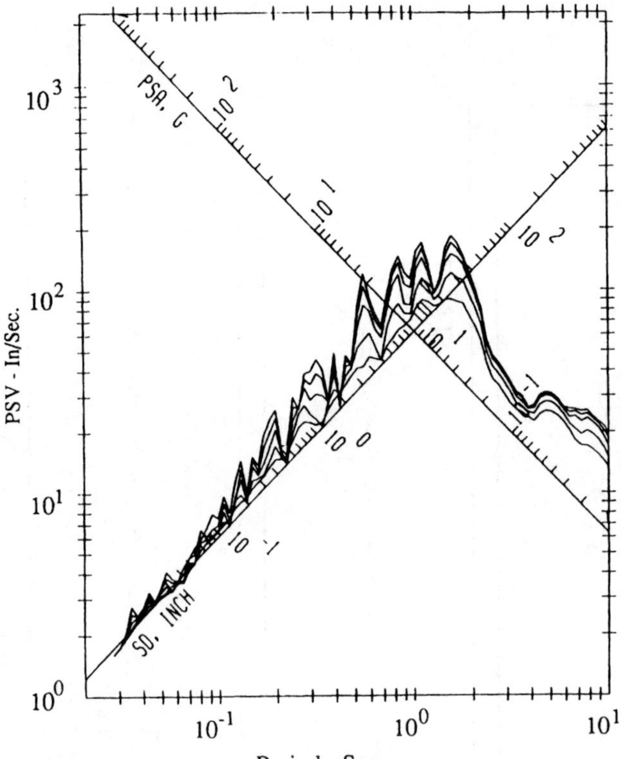

FIGURE 5-3. Response Spectra of the Record Shown in Figure 5-2.

Note: The spectra for damping ratios of 0.005, 0.01, 0.02, 0.05, and 0.10 are shown.

fied is shown in Figure 5-6. This dead-end structure supports the transmission line entering the substation. The pier supporting this tower is 10 m (34 ft) deep, but was still inadequate to prevent the tower from shifting. Figure 5-7 shows a crack in the site that opened about 30 cm (1 foot) because of lateral spreading. The increased separation between equipment increases interaction forces applied by the conductors to bushings or post insulators and can cause damage.

5.3.3 Soil–Structure Interaction

There have been several cases where transformers have rocked back and forth. In some cases, the transformers and their foundation slabs remained tilted after the earthquake. In other cases, the foundation slabs were still level, but damage to connecting bus work indicated soil-structure interaction had occurred. Figure 5-8 shows the foundation slab had moved relative

FIGURE 5-4. Subsidence of the Site Surface Caused the Equipment To Settle.
Note: The bus support structure, which was on piles, did not move; this induced conductor loads on the circuit breaker bushings.

to a sidewalk adjacent to the slab. (Additional material on soil-structure interaction is contained in Appendix B.)

5.4 RECOMMENDED DESIGN CRITERIA FOR SUBSTATIONS

This Manual does not propose any seismic design criteria. For suggested seismic design criteria for substation equipment, refer to the Institute of Electrical and Electronic Engineers (IEEE) Standard 693, *Recommended Practices for Seismic Design of Substations.*[2] The main thrust of Standard 693 is to establish qualification methods and suggest loads for three levels of earthquake hazards. Qualification methods have been tailored to each of the various classes of equipment found in substations. For the seismic design of substation structures, refer to the American Society of Civil Engineers design guide, *Substation Structure Design Guide.*[3]

There is a disadvantage to ordering equipment with special seismic qualification specifications. The qualification of equipment with special seismic specifications using analysis or testing can only be done after the equipment is designed and/or built. There have been several instances where it was learned late in the order cycle that the equipment could not satisfy seismic qualification tests or the quality of the seismic qualification analysis was very poor. Because of the long lead time for delivery of this specialized equip-

FIGURE 5-5. A Lightning Arrester Tilted as a Result of Soil Instability.

ment, seismic specifications have often been waived because new equipment could not be ordered and still meet the construction schedule. The introduction and use of IEEE Standard 693 should largely eliminate this situation.

The anticipated benefits of improved substation seismic design, seismic equipment specifications, and installation practices will be realized only if these practices are diligently used. Seismic design guides have been developed and promulgated with the expectation that they are cost-effective. If these methods are not followed, the cost and disruption associated with earthquake damage will continue. Indeed, if damage is observed when improved methods and equipment are available, the perception will be that more stringent requirements or regulations are needed. Examples of practices that circumvent good earthquake performance include poor station layout that limits conductor flexibility; situations where seismic specifications are not incorporated into equipment purchase orders; and cases where

FIGURE 5-6. A Dead-End Transmission Line Tower at a Substation Tilted Because of Soil Liquefaction.

seismically qualified equipment was not ordered because of a slight increase in cost, where seismic specifications were waived because a construction schedule could not be met, and where as-built facilities were not inspected to substantiate that improved methods were being used.

5.5 COMMON FAILURES

Before describing the equipment found in substations, several common failure modes will be discussed to avoid repetition. The most frequently observed and most serious problem is the failure of porcelain members in high-voltage equipment. Other common failures have involved equipment

FIGURE 5-7. Crack Resulting from Lateral Spreading.

Note: A crack about a foot wide opened up in this site due to lateral spreading. The site was located on poor soil near a river.

anchorage, batteries and battery racks for emergency or station power, and various types of liquid storage tanks. There have also been failures of cast-aluminum hardware used to fabricate conductors and busses.

5.5.1 Failures of Porcelain Members

Because of the electrical properties of porcelain, it is found in virtually all high-voltage power system equipment. Unfortunately, the brittle mechanical properties of porcelain have resulted in failures from earthquake-induced loads. Composite fiberglass-polymer insulators and bushings for

FIGURE 5-8. The Rocking of This Transformer Foundation Slab Was Attributed to Soil-Structure Interaction.

Note: Rocking can cause large deflections at bushing conductor connections.

high-voltage applications are now available to replace porcelain in many applications. The performance of these materials in vibration-table tests has been excellent. The actual earthquake performance of this material in high-voltage applications is limited, but has been good.

Some composites, such as some types of polymer concrete, exhibit brittle characteristics similar to porcelain.

The tensile strength of porcelain is much lower than materials normally used as structural members, such as aluminum and steel. Two other characteristics of porcelain are also troublesome. Most common structural materials are ductile, so that a localized stress will cause local yielding and the concentrated stress is redistributed over a larger volume. This does not happen with porcelain. With porcelain, when the stress at any point within the member exceeds a threshold level, the member will fail catastrophically. Another factor that complicates the use of porcelain and the analysis of failures is that, in the manufacture of porcelain, voids and inclusions are incorporated into the material. These act as local stress raisers, so that a member may fail when the overall stress is relatively low. The uncertainty of the number, size, and distribution of these flaws means that porcelain members exhibit large variations in strength compared to traditional construction materials. There is also a size effect, so that larger members have a higher probability of containing an inclusion of a size that can cause failure than in smaller members.

Porcelain members of equipment operating at 220 kV and above are generally vulnerable to earthquake damage. The higher the voltage, the higher the vulnerability. However, the seismic vulnerability is also related to the equipment design and to practices used to install the equipment. For example, there are 230 kV circuit breakers that have failed at peak ground accelerations around 0.05*g*, and other circuit breakers of a different design and operating at 500 kV that have survived ground accelerations in excess of 0.5*g*. Generally the performance of equipment operating at 115 kV and below has been good if it is properly anchored and the conductor is configured properly. Equipment operating in the range between 115 kV and 230 kV has had very limited seismic exposure, so its performance is not known. The earthquake performance of equipment in the 765-kV class is unknown, but is expected to be at least as vulnerable as the 500-kV equipment.

Porcelain members are usually configured as vertical columns with electrical connections or masses supported at their upper ends. During earthquake vibrations, this places transverse loads at the top of the column, which induces large moments at the base of the column. The columns are assembled in one of two ways. The first type of construction has a metal flange or cap bonded to the end of the porcelain member, as used, for example, for lightning arresters, post insulators, and some bushings. In this case, the column is a cantilever beam, and the large moment at the lower end of the beam develops into high stresses in the outer shell at the base.

The other type of assembly has internal members that are in tension (tendons) to clamp the porcelain members together, as, for example, in some types of bushings and support columns on some live-tank circuit breakers. The porcelain members, which are usually pressurized with an insulating gas, are sealed with a gasket. In this case, the large moment at the base of the column may reduce the compressive load on one side of the gasket seal and allow the internal pressure to leak out or to blow out the gasket. A loss of insulating gas may result in flashover, or the activation of an interlock that prevents the device from operating. A blown gasket will allow porcelain-to-porcelain or porcelain-to-metal contact, which can create a stress concentration leading to porcelain failure.

The seismic design and installation of equipment require knowledge of the seismic loads induced in the member. Three basic loadings are possible:

- inertial;
- imposed loads caused by relative deflections of connected equipment; and
- impact-induced loads.

Each has several factors which influence porcelain performance.

A piece of equipment that moves when the ground moves will experience inertial forces. The ground accelerations can be amplified by the dynamic response of the equipment, by the dynamic response of the equipment support, and, if mounted in a structure, by the dynamic response of the structure. The inertial forces are distributed over the equipment and will be proportional to the mass and acceleration. Methods of equipment analysis are discussed in IEEE Standard 693. Most substation equipment exhibits low damping, and a damping factor of 2% is suggested for analysis unless higher damping can be demonstrated. Porcelain members, such as bushings, are typically near the highest point of an item of equipment, so they will experience the largest amplification of ground motion and thus the largest inertial loads.

A second type of load on porcelain members may be more severe than inertial loads. Substation equipment is usually interconnected by electrical power conductors connected to bushings. Loads arise from the interaction between adjacent equipment items or parts of the same equipment. Thus, interaction loads are applied to the bushings. The interaction loads are significantly influenced by the flexibility and slack of the conductor interconnections.

Conductors can be "flexible" cable or "rigid" aluminum pipe or I-beams. A flexible conductor, or *strain bus,* is fabricated from twisted aluminum conductor. In transmission substations, it is typically more than 1.5 in. in diameter, has a length less than several feet, and is relatively stiff. A rigid conductor is fabricated from aluminum pipe or I-beam and, if installed with dog-leg bends, can be relatively flexible. Either type of conductor can transmit large forces unless special provisions are made to decouple them. Consider a current transformer and an adjacent live-tank circuit breaker. Each piece of equipment is mounted on a support structure that may be 4 ft high, and the equipment may be 20 ft tall. Flexibility in the support structures and the equipment, particularly if the equipment can rock, will result in large deflections at the top of the equipment where the conductor connects the two pieces of equipment. If the slack in the electrical connection between the equipment cannot accommodate the combined peak deflections of both pieces of equipment, the conductor can exert a force on the porcelain that can shatter the post insulator or bushing.

In addition to the above interaction loads, there are other potential load sources that affect porcelain members of power equipment. There can be interaction within a single piece of equipment. For example, multihead, live-tank circuit breakers may have interaction between interrupter heads supported on different columns. Another example is loads applied by long flexible conductors due to the dynamic response of the conductor or its support. Also, when equipment moves because of inadequate anchorage, such as in the sliding of a transformer, electrical connections to the equipment may be loaded. All of these loads are applied to the bushings or other porcelain members and can contribute to their failure.

The third type of load, nonlinear and impact-induced loads, can also significantly influence the performance of porcelain members. Even minor impacting can create very large loads that can cause catastrophic failure of porcelain members. The source of the impacting can be difficult to identify. Nonlinear forces can be introduced as slack in flexible conductor is drawn taut. A similar effect occurs when an object supported by an insulator string or with guys experiences an earthquake. Impact loads are also associated with anchorage and include lack of shims under an anchor point, damaged or missing grout under an anchor point, loose anchor bolts, or anchor bolts in oversize holes. When any of these conditions exist, impacting at the anchor points can occur at relatively small earthquake motions.

Another source of impacting is in equipment support structures that use bolted connections with slots rather than holes. Under seismic loads, even tight bolts may allow bolted members to slide, and when the bolt reaches the end of the slot, a severe impact is imparted to the entire structure. Tendon or spring-loaded porcelain members tend to impact when the seismic load exceeds the preload. This can be very severe when gaskets are blown out, so that a porcelain member will come in direct contact with another porcelain member or metal gasket seat. In addition to creating large loads, impacting and nonlinear loads can excite higher modes of vibration of the equipment.

Examples to illustrate some of the failures discussed above are given in the equipment sections that follow.

5.5.2 Failures of Equipment Anchorage

Some of the observed power system damage can be attributed to the lack or failure of equipment anchorage. Inadequate anchorage allows equipment to move or fall over. For equipment that moves, a lack of flexibility or slack in connections can cause the connections to break or load the equipment so that it fails. Examples are failed piping, broken conductor connections, fractured bushings, and fractured post insulators. The failure of an anchorage, per se, usually does not cause the equipment to malfunction, unless the equipment slides or falls over and is thereby damaged.

It should be emphasized that anchorage, as used here, refers to the entire load path from the equipment to its foundation and must take into account the stiffness of the support structure, the equipment near anchorage points, and stiffness at the anchor bolts or welds. A flexible anchorage may allow rocking of the equipment, which can cause large deflections at conductor connections near the top of the equipment. Another problem with flexible anchorage is that it may lower the natural frequency of the equipment so that it is more in tune with the high-energy frequency range of the earthquake (0.5 Hz to 6 Hz). Similarly, flexible support structures and connection details in components mounted to large equipment can cause amplified response and damage. Detailed analyses of equipment damage frequently find that the flexibility of the equipment and its support structure are often the cause of the damage.

Several issues should be considered in anchoring switchyard and control house equipment. The anchor bolt embedment depth should be sufficient to force failures to occur in the bolt rather than in the concrete, as this will dissipate more energy and should eliminate brittle failure. Mechanical anchors used for heavy switchyard equipment should be specified by the utility and selected from those recommended by the anchor manufacturer as appropriate for dynamic loads. High-strength bolts (generally considered ASTM A325 and above) are not recommended, as brittle failures are to be avoided. In retrofitting equipment, high-strength bolts may be needed, but many engineers are reluctant to use grades with strength of ASTM A490 or above. Anchor strength ratings used to meet anchorage specifications must be reduced to account for the spacing of the bolt pattern used. Expansion anchors may be used for control room equipment racks, but the minimum size bolt should generally be ½-in. diameter. The factor of safety should meet building codes (typically a factor of safety of 4). Bolts should pass through a structural member in the equipment framing. When only sheet metal is available, heavy plate washers should be used to prevent tearing of the sheet metal. It is important that the as-built embedment length meet manufacturer's test conditions in determining bolt strength. Of particular concern are situations where anchor bolts pass through inverted channels used in the equipment framing system, which causes bending in the bolts and reduces embedment length for some types of anchor bolts. The placement of anchors and the equipment configuration should avoid prying action on bolts or welds. Proper installation of anchors, particularly hole size and hole preparation, is needed to meet manufacturer's strength ratings.

For bolted connections, the bolt hole should be appropriate to the bolt diameter, and oversize holes should be avoided. An oversize hole will allow the equipment to slide, and, when the equipment strikes the bolts, large impact stresses will be induced in the supported structure and in the bolt. Equipment manufacturers sometimes provide oversize bolt holes to allow for misalignment in placing cast-in-place bolts. This can be mitigated by placing a thick, close fitting washer under the nut and welding it to the equipment, as illustrated in Figure 5-9. This subjects the bolt to bending stresses. If the bolt is approximately centered in the oversize hole, a cylindrical spacer can be slipped over the bolt to eliminate the large gap, or, if the base plate is grouted, the enlarged hole can be grouted so that the welded heavy washer is not needed.

Elements in the load path, including the equipment and its support structure, should be stiff. Anchorage design should consider shear, vertical accelerations, overturning moments, prying action, and torsion due to eccentricities. If the equipment is qualified using IEEE 693, then the Seismic Outline Drawing, which contains anchorage specifications, should be consulted.

An anchorage that allows the equipment to slide or rock can cause impact loads that can damage porcelain members. Impact loads may also cause

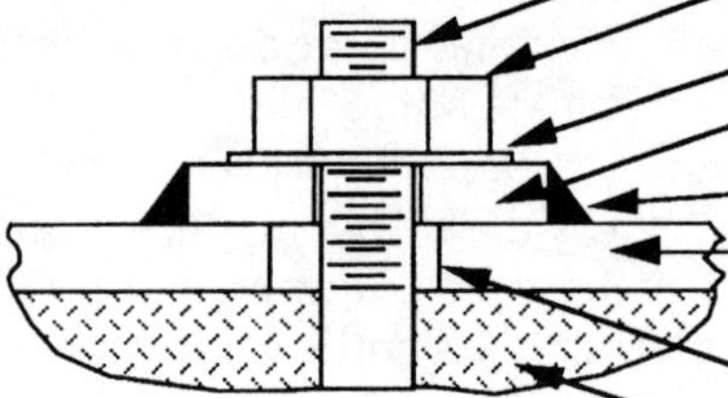

FIGURE 5-9. Thick Washer Welded to Equipment To Eliminate Effects of Oversize Bolt Holes.

relays within the equipment to chatter; this may cause undesired equipment actions.

5.5.2.1 Friction Clips. Friction clips have been used to anchor switchyard equipment, as well as cabinets and consoles in control houses. They are often used because of their convenience. Often, anchor bolt patterns are not known when foundation slabs are poured, so cast-in-place bolts cannot be used. Also, the use of cast-in-place bolts requires that the equipment be lifted over the bolts rather than sliding it onto the foundation slab. Friction clips, however, can be vulnerable in earthquakes. The earthquake performance and upgrading of friction clips used in switchyards are discussed in Section 5.12, Circuit Breakers. The use of friction clips on cabinets and control consoles is discussed in the Section 8.5.1, Control Console and Status Board.

5.5.3 Failure of Cast-Aluminum Hardware

Cast-aluminum hardware, which is used in equipment and to support equipment, conductors, and busses, has experienced brittle failures. There have also been failures due to the deformation of cast-aluminum members used in disconnect switches. Examples of these failures will be illustrated in the specific equipment sections below.

5.6 SUBSTATION BUSSES, CONDUCTORS, AND THEIR SUPPORTS

Most substations use power conductors in three ways. Transmission lines usually enter the substation from a dead-end transmission tower. Within the substation, there are busses and bus support structures. Conductors also connect power equipment to each other and to the busses.

5.6.1 Dead-End Transmission Towers

The last transmission tower before a transmission circuit enters a substation is usually a dead-end structure—that is, each line terminates at an insulator string on the structure. The structure is typically an A-frame, fabricated from steel members, a lattice structure, or a frame fabricated from plates to form box beams. A schematic diagram of a dead-end structure where it enters a substation is given in Figure 5-10.

5.6.1.1 Earthquake Performance of Dead-End Structures. While the earthquake performance of transmission towers has generally been good, there have been several failures in recent earthquakes. Possibly the most spectacular was associated with soil liquefaction and the failure of one leg of a dead-end tower at a substation. The tower was fabricated from plates and supported on a concrete pier that extended 10.3 m (34 ft) below grade. Figure 5-6 shows an overview of the tower, which can be seen tilting to the left with a tear in the leg nearest the camera. A close-up view shows the opposite side of the leg (Figure 5-11).

Two other types of failure have been observed on dead-end structures. In one, a hook or bolt head in a slotted connection that attached the insulator string to the structure became disconnected and allowed the line to drop. In another dead-end structure, the lines attached to each side of the structure were connected with an aluminum-pipe conductor that passed over the top

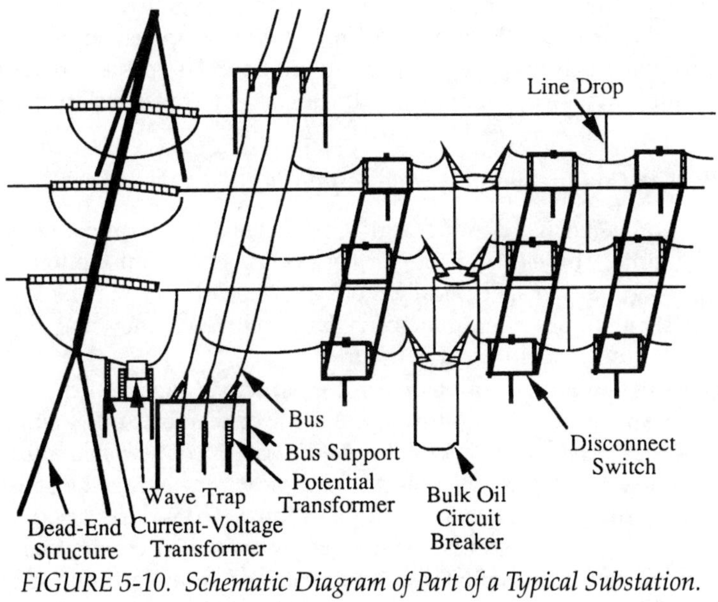

FIGURE 5-10. Schematic Diagram of Part of a Typical Substation.

FIGURE 5-11. Close-up View of the Damaged Transmission Tower Shown in Figure 5-6.

Note: The material used to fabricate the transmission tower failed in what appeared to be a brittle fracture.

of the structure to connect the lines on each side (Figure 5-12). A cast-aluminum fitting at the end of the pipe failed, allowing the pipe to drop and short the line to the structure (Figure 5-13). The dynamic response of the tower to the earthquake can amplify the ground motion. There have been other tower failures, which are discussed in the section on transmission towers.

5.6.2 Busses, Conductors, and Their Supports

In general, a *bus* is a conductor that provides a common connection to several circuits. The electrical power connections between equipment are referred to as *conductors*. Physically, a bus and a conductor may be identical, although since busses often must carry larger currents, they may be larger than the conductors connected to them.

Utilities use two types of busses and conductors: rigid or flexible. Rigid conductor is made of aluminum pipe 2–8 in. in diameter or of aluminum structural sections. Flexible conductor is fabricated from twisted aluminum cables. Each system has it advantages and disadvantages. It is the consensus of substation design engineers that, for earthquake considerations, properly installed flexible conductors have advantages over rigid conductors. However, this has not been definitively demonstrated from earthquake performance. Rigid busses may have advantages for wind loading, reduced phase-

FIGURE 5-12. Overview of Rigid Conductor Passing over Dead-End Bus Support Structure.

FIGURE 5-13. Failed Cast-Aluminum Fitting at the End of the Rigid Conductor in Figure 5-12.

to-phase spacing, and lower profile, because sag does not affect conductor-to-ground clearance. The use of a given system may be influenced by historical practices and the desire to maintain uniform practices throughout a utility.

5.6.2.1 Earthquake Performance of Bus and Conductor. Failures of rigid and flexible conductors runs have been observed. Figure 5-14 shows damaged post insulators that caused the failure of a long run of flexible conductor. Figure 5-15 shows one phase of a rigid bus that failed. In this case, the failure was primarily attributed to the failure of cast-aluminum hardware used to connect the conductor to the top of the post insulator. In both of these cases, the effects of equipment interaction with the bus did not contribute to the failures.

A distinction must be made between the flexibility of a conductor, the flexibility of a connection, and slack in a connection. A conductor can be fab-

FIGURE 5-14. Post Insulators Supporting Flexible Conductor Failed.

FIGURE 5-15. One Phase of a Rigid Bus That Failed.

Note: This phase of a 500-kV bus fell, in part, because of the failure of cast-aluminum hardware used to connect the bus to the post insulators.

ricated from fine braided wire, so that it can be easily flexed by hand; however, if it is used to connect equipment and is not provided with slack, the connection will not be truly flexible because it can become taut and apply interaction loads. Likewise, a long run of rigid bus that has a dog-leg bend in it can provide a very flexible connection.

The flexibility of conductor connections between substation equipment appears to affect earthquake equipment damage. The effects of conductor installation practice for specific types of equipment will be discussed in the equipment sections that follow. While it is relatively easy to install a flexible conductor so that it has adequate slack and flexibility, good practices are not always followed. Often, the configuration of the conductors is not engineered, but left to the field crews to install according to traditional practices, which unfortunately may be inappropriate for earthquakes. Figure 5-16 shows conductors between circuit breakers and adjacent disconnect switches. This shape provides a flexible connection and does not compromise ground-to-conductor clearance. Figure 5-17 shows another position in the same switchyard. In this case, the lack of slack probably contributed to the failure of the disconnect switch.

The sequence of failures during an earthquake has not been directly observed but can sometimes be deduced. Since equipment is connected together by conductors, the failure of one item may cause a cascade of fail-

FIGURE 5-16. The Configuration of the Conductor between the Circuit Breaker and the Disconnect Switch Provides a Flexible Connection.

ures as one item damages an adjacent item. After the fact, it is often impossible to ascertain whether failures are due to the cascade effect or whether individual items would have failed even if the adjacent item was undamaged. Occasionally, by a careful review of damage patterns within the entire switchyard, initiators of damage can be identified. Cascading failures might be prevented with the introduction of breakaway connectors. That is, connections could serve as a mechanical fuse, so that if the load exceeded a predetermined value, the connector would fail and thus limit the load that could be applied through the connection. As will be seen in subsequent sections, there have been nonengineered connector failures. Frequently, the connector would fail when an item of equipment would tip over. Unfortunately, this would usually occur after the adjacent item was also damaged. Breakaway connections are discussed below in Section 5.6.4 on mitigation methods.

FIGURE 5-17. Configuration of Conductor without Slack in Flexible Connection.
Note: The lack of slack in the flexible conductor between the circuit breaker and the disconnect switch probably contributed to the failure of the disconnect switch.

Vertical drops from overhead conductors to equipment are needed in both flexible and rigid conductor systems. A line drop is illustrated in Figure 5-10. The line drop can be from a flexible conductor, from a rigid conductor, or from a conductor support structure. There have been several failures of vertical drops. On a rigid conductor, the field weld attaching the drop to the overhead conductor failed. Clearly, a falling rigid conductor can damage nearby porcelain. There have been several failures associated with flexible conductor drops; these failures are not well understood. It is not clear whether the dynamic response of the drop line or the overhead conductor contributed to the damage. There has also been damage to drop lines from overhead conductor support structures. Again, the dynamics are not well understood, but the vertical dynamic response of the support structure may have contributed to the damage. This will be discussed further in Section 5.15, Wave Traps.

The vulnerability of cast-aluminum hardware was noted above. The ends of conductors often have this type of hardware, the failure of which is shown in Figure 5-18.

Vertical drops to equipment at sites that may be subject to soil liquefaction or subsidence present special problems. These are discussed in Section 5.12, Circuit Breakers.

Special considerations for conductor connections to suspended equipment and lightning arresters are discussed in subsequent sections.

FIGURE 5-18. Cast-Aluminum Fittings That Form the Termination of Flexible Conductors Exhibiting Brittle Failures.

5.6.3 Bus and Conductor Support Structures

Several types of structures are used to support busses and conductors. The simplest is a tubular vertical column topped by a post insulator and hardware to support the conductor. Simple frames are used that are constructed from structural shapes or lattice. Some utilities use much larger frames that also support disconnect switches.

5.6.3.1 Earthquake Performance of Bus and Conductor Support Structures. Tubular columns support conductors and other equipment in substations. There have been failures in the weld connecting the column to the base plate. This type of failure is discussed in Section 5.9, Lightning Arresters.

The conductor support structures shown in Figure 5-19 had several weld failures. While the structures did not collapse, several insulator supports failed, dropping their conductors and shorting them to the support structure.

Figure 5-20 shows the base of a large conductor support structure. The 3-in.-thick base plate of the structure had a vertical crack starting at the weld that joins the plate to the column and extending through the base plate. The crack exhibited brittle failure characteristics.

The vulnerability of the cast-aluminum hardware used to support conductors has been noted above. Figures 5.21 and 5.22 show, respectively, failed hardware supporting a 500-kV conductor and cast-aluminum hooks used to support a 34.5-kV bus.

Cast-aluminum hardware used to support flexible conductor has also failed, as shown in Figure 5-23.

FIGURE 5-19. Conductor Support Structures Fabricated from Plate Had Weld Failures That Contributed to Insulator and Line Damage.

5.6.4 Mitigation and Retrofit of Substation Busses, Conductors, and Their Supports

The flexibility of connections can be improved by providing increased slack to flexible conductor connections. With a review of the system, some conductors may be able to be moved and some replaced, rather than replacing all conductors that have inadequate slack. In the case of rigid bus, the procedure suggested for new construction given in Section 5.6.6, Recommended Installation Practices for Substation Busses, Conductors and Their Supports, may be applicable.

In an earthquake, the flexibility of the connection must be sufficient to accommodate the relative movements of the conductor end points. Thus, the peak deflections of each connection point must be conservatively estimated and added together to determine the total relative movement. An additional factor of safety should also be applied. Section 6.9.2, Observed Component Displacements, of IEEE Standard 693 provides guidance for

FIGURE 5-20. Thick Base Plate of a Conductor Support Structure with a Crack through the Plate.

FIGURE 5-21. Cast-Aluminum Hardware Supporting a 500-kV Rigid Conductor Failed.

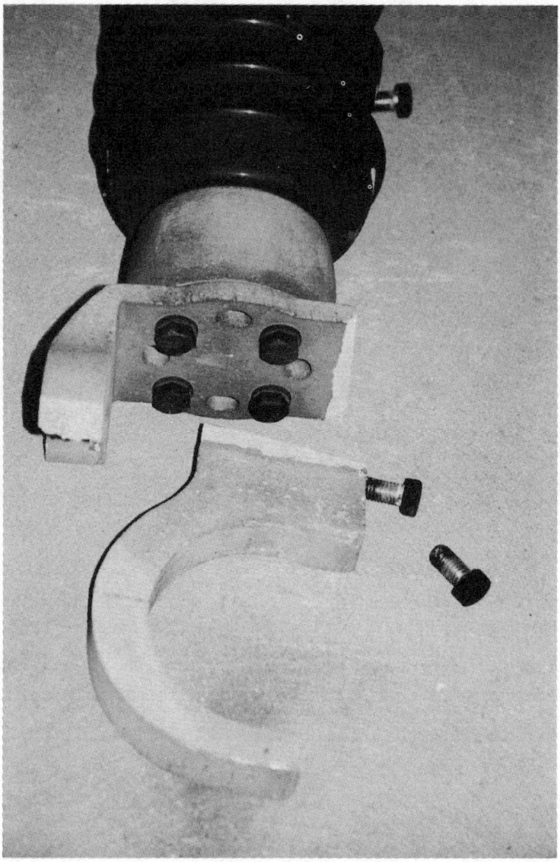

FIGURE 5-22. Cast-Aluminum Hooks Used To Hold a 34.5-kV Rigid Bus Failed.

deflections to be expected for different voltage equipment of different natural frequencies. This guidance is reproduced in Table 5-1. The definitions of high-, medium-, and low-frequency equipment are given in Table 5-2. Standard 693 recommends a margin of safety of 50% be applied to the total expected relative movement. Unfortunately, electrical design considerations may limit the flexibility of connections, because of short circuit loads and the need to maintain minimum phase-to-phase or phase-to-ground separation. While new designs may be able to accommodate both criteria, this may be at the expense of a less compact substation design.

The methods for implementing connection flexibility provisions for flexible and rigid conductors can affect earthquake-induced loads. For flexible conductors, as conductor connection points move toward each other, the conductor will bow and exert relatively small loads on the connection points. However, as the connection points separate, the slack in the conduc-

FIGURE 5-23. Cast-Aluminum Hardware Used To Support Flexible Conductor Has Failed.

TABLE 5-1. Typical Equipment Displacements (mm).

Frequency	138 kV	230 kV	500 kV
High	25–50 (1–2 inch)	25–75 (1–3 inch)	100–300 (4–12 inch)
Medium	50–150 (2–6 inch)	70–200 (3–8 inch)	200–600 (8–24 inch)
Low	150–500 (6–20 inch)	200–1,000 (8–39 inch)	300–1,500 (12–59 inch)

Source: Endnote 2.

TABLE 5-2. Frequency Definitions and Representative Equipment.

Frequency	Definition: Representative Equipment
High	8 Hz and greater: transformer, tank reactors, dead-tank circuit breakers
Medium	2.5–8 Hz: disconnect switches, live-tank circuit breakers, capacitor banks
Low	under 2.5 Hz: capacitor voltage transformers, current transformers, wave traps, suspended components

Source: Endnote 2.

tor will be reduced, and, as it draws taut, the stiffness of the connection will rapidly increase so that an impact load may be imposed by the conductor on the connection points. Thus, if phase-to-phase separation and ground clearance do not impose restrictions, generous provisions for slack should be made. Connection flexibility for a rigid bus is usually provided because of the need to accommodate thermal expansion of the bus. This can be provided by short segments of flexible braided conductor, many layers of thin aluminum sheet formed into a half circle, or telescoping conductor sections. In general, these provisions for thermal expansion will be inadequate for the size of relative motions expected in earthquakes. Also, the telescoping connection only provides flexibility along the axis of the conductor, and the half circle tends to be stiff in the transverse direction. Moderate-length rigid bus sections with dog-legs can accommodate large relative deflections and place relatively small loads on end connections (Figure 5-24).

It is vital that mechanical strength specifications for bus and conductor support hardware be established that take into account earthquake loads and that these specifications are satisfied. The brittle failure of cast-alumi-

FIGURE 5-24. Example of a Gas-Insulated Circuit Breaker with Dog-Leg Connection on a Rigid Conductor To Provide a Flexible Connection.

num hardware used for conductor supports and connections has been very common. It is not known whether these failures have been due to an underestimation of earthquake loads, to an inadequate factor of safety for the brittle character of the material, or to poor designs with stress concentrations that cause premature failure. Another possibility is that the electrical currents modified the mechanical properties of the materials.

In general, the separation of equipment should be as small as practical because less conductor is needed to span the gap and the seismic loads on porcelain members supporting the conductor are reduced. Excessive spacing may also facilitate amplified dynamic response of the conductor, which increases support loads on porcelain members. Closely spaced equipment has the disadvantage of providing less working space around equipment and can lead to potential problems if replacement equipment is larger than the preexisting equipment.

Some utilities that use rigid conductor have introduced flexible links between the bus and the equipment. Figure 5-25 illustrates two configurations using this approach. In Case A, if the supports for the rigid conductor fail, the conductor could fall and damage the bushing. In Case B, the rigid conductor does not extend over the bushing, so the failure of its support is less likely to damage the bushing.

Failures of vertical drops from an overhead conductor to equipment were noted. For sites with the potential for subsidence, a rigid bus with a rigid drop provides little slack to accommodate relative vertical movement. The

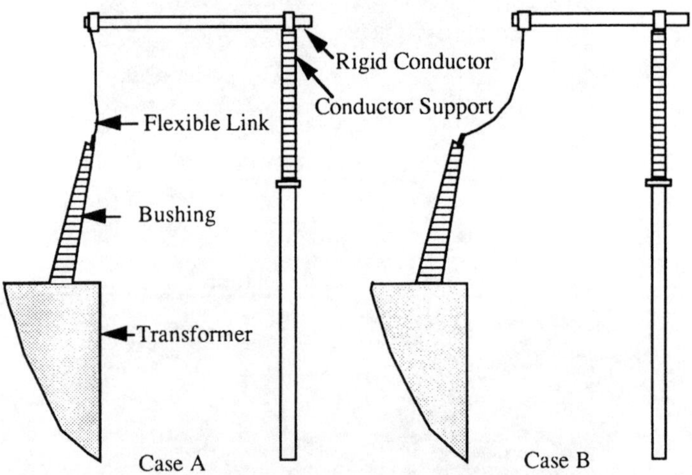

FIGURE 5-25. Flexible Drops from Rigid Conductor to Equipment Can Provide the Advantages of Both Rigid and Flexible Conductors.

Note: Slight changes in configuration can also reduce the risk of earthquake damage to critical equipment.

introduction of a horizontal offset at the base of the drop to the equipment would provide some vertical flexibility. Even if the site does not experience subsidence, flexibility in the vertical drop could reduce loads on the connection resulting from the vertical dynamic response of the overhead conductor. Providing slack to the vertical drop of a flexible conductor can create problems with minimum phase-to-phase separations. Relatively small increases in the length of the vertical conductor will allow relatively large lateral displacements (Figure 5-26). Shifting the position of the upper connection of the vertical drop of the center phase may allow phase-to-phase separations to be maintained.

The effect of nonengineered breakaway connections was discussed above. Several engineered breakaway connections have been proposed, but are not now commercially available. Incorporating these devices could provide several benefits, particularly for retrofitting existing facilities. Their design would have to be such that they could be easily incorporated into existing configurations. A properly designed breakaway connection would limit the loads placed on porcelain members, such as bushings and post

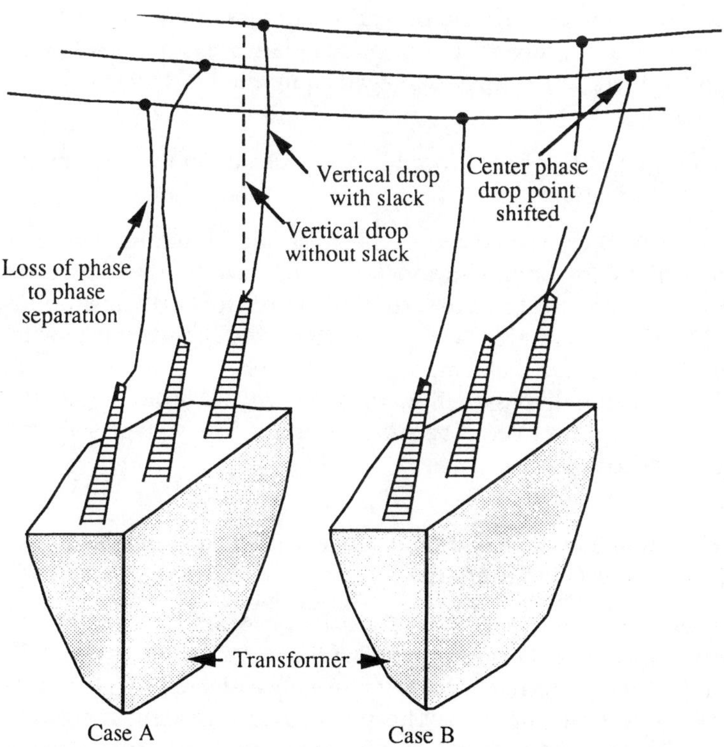

FIGURE 5-26. Alternatives for Vertical Drops Using Flexible Conductor.

insulators, because of relative motion between equipment and inadequate conductor flexibility.

Two types of breakaway connectors have been proposed. The first is an engineered version of the conductor connection that, when it fails, breaks the electrical connection and mechanical support at the connection. The second type would allow the normal mechanical connection between the conductor and the equipment to fail, but would retain a flexible electrical connection. Each of these approaches has advantages and disadvantages. The first design has an advantage if an item of equipment falls over. Since the entire connection between the equipment separates in this case, all interaction loads are eliminated. However, in some configurations, the loss of mechanical support at the breakaway connection will allow the unsupported conductor to fall, and it could short against other equipment and cause damage or create a hazard for personnel in the switchyard. If the separation between equipment is kept small, this would not be a problem, as length of the unsupported conductor could not reach a grounded component. The second type of breakaway connector has the advantage that, if interaction breaks the mechanical connection and the equipment is not damaged, electrical continuity would be maintained and service would not be disrupted. However, if item of equipment falls, the remaining electrical connection would probably not have adequate slack, and loads would be transmitted to the adjacent, undamaged equipment.

5.6.5 Emergency Response Procedures for Substation Busses, Conductors, and Their Supports

There have been several cases where cast-aluminum hardware used to support conductors, both rigid and flexible, has failed. To quickly restore service, rope or wire were used to secure the conductors to their supports (Figure 5-27). There have also been cases where the high-voltage switchyard damage was so severe that temporary conductors were installed; these temporary conductors bypassed the entire switchyard to undamaged transformers so that power could be supplied to the low-voltage switchyard to maintain service.

5.6.6 Recommended Installation Practices for Substation Busses, Conductors, and Their Supports

Improved earthquake response can be obtained by providing flexible connections, whether flexible or rigid conductors are used. In general, a flexible conductor is preferred because it is usually easier to provide flexible connections between equipment. The procedures regarding mitigation suggested in Section 5.6.4 are applicable for retrofitting and new construction.

FIGURE 5-27. A Rope Was Used To Secure a Rigid Conductor to the Remains of a Cast-Aluminum Hook That Holds the Conductor to the Post Insulator.

5.7 POWER TRANSFORMERS

Power transformers perform the vital function of transferring power between circuits operating at different voltages. They also may contain tertiary windings that generate voltages that are used for station power and other purposes.

Transformers are constructed in one of two ways. In a core-type transformer, the transformer coils have a framing system supporting the core within the transformer case. In a shell-type transformer, the coils are form-fit into the case. The shell-type transformers typically have a smaller footprint and a larger height-to-width ratio. It has been claimed that the internals of shell-type transformers are more seismically rugged than core-type transformers, but there is no documentation from earthquake damage to support the view. There have been internal transformer failures; however, the causes of the damage have not been documented. After the Northridge earthquake, an evaluation of transformer performance suggests that there may be an accelerated degradation of electrical performance because of the shifting of the core and an increase in internal electric fields that deteriorate insulation. The electrical degradation appears to take place over several years, so that the causal relationship to earthquakes is difficult to trace.

Transformers have several components that are related to their earthquake performance: sudden pressure and protective relays, anchorage, radiators, bushings, conservators, lightning arresters, tertiary bushings, and surge arresters. Lightning arresters are discussed in Section 5.9. Some transformer installations have transfer busses so that a spare transformer can quickly replace a unit that must be taken out of service. As these bus configurations are unique for use with transformers, they will be discussed in the transformer section.

The effect of the loss of function of a transformer is generally significant, unless a spare transformer is available or there is a second transformer bank in parallel with the damaged unit. The consequences of transformer damage will depend on system configuration and other system elements that may be damaged or out of service.

The emphasis of this section is on bulk power transformers; however, distribution transformers are briefly discussed.

5.7.1 Sudden Pressure, Bucholtz and Protective Relays

Sudden pressure and Bucholtz relays are meant to limit damage should there be an internal transformer fault. When there is an internal fault, like arcing between adjacent coils, gases are generated that cause the transformer internal pressure to increase and oil to move into the conservator. The pressure increase or movement of oil is sensed and causes the transformer to be isolated by the opening of circuit breakers on each side of the transformer. Earthquake-induced vibrations can also cause changes in pressure that can trip the sudden pressure relays. Mercury switches have been used in sudden pressure systems, and these were easily tripped by earthquake vibrations. To avoid unintended tripping, at least one of the major utilities has replaced mercury-based switches.

Typically, sudden pressure relays must be manually reset. In any event, the substation will be inspected for earthquake damage. If no damage is observed and shaking levels were low to moderate, the transformer will usually be put back into service. If there is any concern that there was an internal fault, a high-potential test of the transformer will be performed before it is put back into service. The tripping of sudden pressure relays is fairly common, and California utilities respond appropriately. However, in a small earthquake in Kentucky, a sudden pressure relay tripped a transformer off line. This transformer served to connect a nuclear-generating station to the grid. Because of unfamiliarity with the response of sudden pressure relays in earthquakes, it was eight hours before the transformer and power-generating station were put back on line. Make-up power during the disruption reportedly cost about $1,000,000.

Power transformers typically have several protective relays that monitor performance and provide protection. Some protective relays may be activated by an earthquake and cause the transformer to be taken off line. There have been cases where relay lockout flags had been tripped, but it appeared that vibration caused the flag to drop without actually tripping the relay. The false tripping of protective relays should not cause any damage to facilities and should only cause brief, if any, service disruption.

While the vibration-induced action of sudden pressure and protective relays can cause unnecessary disruption of service, there are indications that it can prevent damage to the distribution system in an earthquake. Earthquake-induced vibrations cause distribution lines to swing, and adjacent phases can come in contact. This can cause the lines to wrap around each other, to burn down, or to blow fuses and trip circuit breakers. Damage to the distribution system in earthquakes, most of which is caused by lines coming together, has been compared to that due to a severe winter storm. If the lines are not energized, most of these effects will be avoided. Purposely shutting down the distribution system in an earthquake for a minute or two when an earthquake is sensed could reduce distribution system damage. However, because of the reduced reliability from disruptions associated with the many earthquakes that would not cause distribution system damage, this mitigation measure has not been implemented. If significant load is disrupted, the imbalance of power generation and demand may cause generating units to trip off line, and there can be delays in getting them back on line.

5.7.2 Anchorage

Anchorage, as used in this document, is given a broad interpretation. In addition to the traditional issues of attaching the transformer to its foundation slab, issues related to the foundation slab are also considered.

Four general approaches are used to install power transformers. When little consideration was given to earthquakes, a transformer was set down unanchored on a concrete slab or on rails. Two other approaches address earthquake issues by bolting the transformer to its concrete foundation slab or by welding it to steel embedded in the concrete foundation slab. Within each approach, there are variations. For example, one large utility supports many of its older transformers on a series of closely spaced rails. This allows air to circulate under the transformer to avoid corrosion and provide additional cooling.

There are many transformer configurations and constraints in designing transformer anchorage retrofits. Limitations on cost and materials, as well as geometric constraints, may yield less than optimal designs that are nonetheless adequate.

5.7.2.1 Transformer Foundations. Within California, there has been wide variation in the approach to designing a transformer foundation slab. The most striking difference is reflected in the thickness of the slab, with the thin parts of some slabs only 23 cm (9 in.) thick and some slabs more than 1 m (3 ft) thick. Even very thin slabs with moderate reinforcing have not failed in earthquakes. However, it is difficult to design proper anchors when the slab is so thin. There have been cases were thin slabs have cracked during the process of moving the transformer onto the slab.

5.7.2.1.1 Earthquake Performance of Transformer Foundations. Most foundation slabs have remained level after earthquakes; however, there have been exceptions. In general, the permanent tilt of a foundation slab would be associated with soil liquefaction or differential settlement. However, slab tilt has been observed when there was no indication of soil liquefaction and differential settlement from subsidence seemed unlikely. In the Northridge earthquake, post insulators supporting a low-voltage bus on several transformer installations were damaged. The condition of the soil around the edge of the slab on some installations suggested that the slab had rocked back and forth during the earthquake, although the slab was level. Some slabs were tilted. The rocking of the transformer and foundation slab associated with soil-structure interaction can increase the inertial loads on bushings, lightning arresters, and radiators. The rocking also increases displacements of points where conductors are attached to the transformer bushings.

A gap around the boundary between the soil and the foundation slab was observed after the earthquake. This has been attributed to the motion of the transformer and slab compressing the soil near the slab. A 2.4 cm (1 in.) gap is shown in Figure 5-28, but gaps as large as 3 in. have been observed.

5.7.2.1.2 Mitigation and Retrofitting Transformer Foundations. If a transformer foundation has tilted as a result of an earthquake, it can be re-leveled by inserting tapered shims under the transformer.

Several approaches can be taken to reduce the effect of soil-structure interaction on transformer performance. Generous slack can be provided to accommodate the increased demand on flexible conductor connections. Bushings can be qualified to the High Performance Response Spectra (IEEE Standard 693), robust bushings that exceed the High Performance Response Spectra can be used, or spare bushings can be obtained and stored at a safe location or in seismically adequate racks. Because of high analysis and implementation costs, as a last resort, more elaborate foundations can be used to reduce the effects of soil-structure interaction. A cost-benefit analysis should be performed to determine the appropriate approach. Similar approaches should be considered for new construction.

In one situation were the bushings could not be upgraded, the foundation was retrofitted to reduce soil-structure interaction. In this case, holes of about 12 cm (5 in.) in diameter were drilled through the slab at a slight angle away from the transformer until competent soil was reached, about 18 m (60 ft) deep. A cable anchor was grouted at the bottom of each hole, and the

FIGURE 5-28. A 2.4 cm (1 in.) Gap around This Transformer Was Caused by the Back and Forth Motion of the Transformer Compacting the Soil.

cable was post-tensioned to about 200,000 pounds. In another case, piers were installed adjacent to and tied into the slab. Four to six 60-cm (2-ft) diameter piers up to 6 m (20 ft) deep have been used.

5.7.2.2 Unanchored Transformers. *5.7.2.2.1 Earthquake Performance of Substation Transformers.* Some transformers are installed on a slab without anchorage, but this option should be rejected because of the large direct and indirect cost associated with transformer damage. Utilities that have evaluated the seismic risk associated with inadequately anchored transformers have found that the potential losses justified improved anchorage through retrofitting. Clearly, if retrofitting is justified on existing installations, then new transformer installations, where the incremental cost of anchoring is smaller, should be securely anchored.

The disadvantage to not anchoring a transformer is that the transformer can move large distances relative to the ground. Figure 5-29 shows a transformer that has moved about 60 cm (2 ft). As a result, slack provided to the flexible power conductors was inadequate, and one lightning arrester mounted on the transformer broke. Fortunately, the connections to the other lightning arresters and bushings failed, rather than breaking the porcelain members. If the slab tilts, even larger movement can be expected of an unanchored transformer.

A Japanese utility placed 18-mm (¾-in.) thick, rubberlike mats under unanchored transformers to reduce noise. This seemingly minor modification, a type of base-isolation, had an adverse effect on transformer response and performance during an earthquake. The transformer did not slide during the 1978 Sendai, Japan, earthquake, but many of the bushings were damaged. A postearthquake evaluation conducted by the utility indicated

FIGURE 5-29. The Movement of an Unanchored Transformer.

Note: The unanchored transformer slid about 60 cm (2 ft) in a 6.6 Moment Magnitude earthquake. One lightning arrester failed due to conductor loading. Conductor connections to the other lightning arresters and bushings failed, preventing damage to the porcelain components.

that the mats reduced the natural frequency of the transformers, so that the high-voltage bushings were excited near their resonance and were damaged (Figure 5-30). One or more high-voltage bushings were damaged on each of the five three-phase transformers at the site. The use of the isolation pads was discontinued. This demonstrates that the use of base isolation requires very careful evaluation to assure that the desired effects are achieved. Isolation pads should not be used unless they are included in the seismic qualification of the transformer.

When a transformer moves relative to its support slab, control cables can be damaged (as shown in Figure 5-31). In extreme cases, unanchored transformers can tip over, as happened to the spare transformer shown in Figure 5-32. It is interesting to note that a poor anchor design may contribute to a transformer tipping over. For example, if the anchorage primarily serves to restrain a transformer from lateral movement but does not hold it down, a horizontal acceleration will generate overturning moments, tending to tip the transformer over.

There have been cases of internal damage to transformers. In one case, there were indications of electrical abnormalities immediately after the earthquake, and a high potential test indicated problems. The transformer was removed from the site for repair, but details of the damage are not

FIGURE 5-30. Damage to High-Voltage Bushings on a Transformer.

Note: Two of the three high-voltage bushings on this three-phase transformer failed because of the change in frequency caused by the use of a thin sound-isolation pad placed under the unanchored transformer.

FIGURE 5-31. Unanchored Transformer That Slid and Damaged Control Cables.
Note: Discoloration of the slab indicates that there was also an oil leak.

FIGURE 5-32. Spare Transformer That Was Not Anchored and Tipped.

known. In another case, months after the earthquake, there were electrical abnormalities, and, upon an internal inspection, significant deformations in the coil restraints were observed. Since postearthquake inspections of the transformer internal structural system are rare, it is unknown whether internal structural deformations are common.

5.7.2.2.2 Mitigation and Retrofit of Substation Transformers. Transmission power transformers that serve large metropolitan areas typically range from 180,000 to 500,000 pounds in weight. Large units can exceed 1,000,000 pounds. For a 500,000-pound transformer subjected to 0.5*g* horizontal acceleration, each of four anchor points will have to withstand from 62,500 pounds to 125,000 pounds of horizontal load, depending on the anchorage design. In addition to lateral loads, the anchorage must be able to resist overturning moments and torsional loads due to eccentricities. In designing anchorage, no credit is taken for friction forces.

Three issues must be addressed in anchoring all transformers, be it retrofitting an existing installation or designing a new installation. These issues are evaluating the load path between the transformer and the ground, determining that the strength of all elements in the load path is adequate, and determining that the anchorage is adequately stiff so that excessive movement and impact loads due to deformation can be avoided. For welded connections, the stress distribution across the weld should be evaluated under the various earthquake loads. In retrofitting systems, as will be seen below, load paths can be very complex, so they must be given careful consideration.

Two methods are used to anchor an existing transformer to a concrete slab. This is typically done in such a way that the transformer does not have to be taken out of service or moved from its foundation. The first method welds a plate or bracket to the transformer base that is then bolted to the concrete slab. The second method anchors the transformer through lifting lugs that were part of the original transformer design.

There are several concerns when welding to a transformer for retrofitting. Existing plans for the transformer may not be adequate to indicate good anchor points. The original design may not accommodate the concentrated anchor loads that will be experienced in an earthquake. In welding directly to the transformer tank, significant heat will be generated, and it may not be clear whether an internal coating, components, or oil will be damaged. There are also safety issues associated with welding on a transformer tank. While the use of small welds can reduce heating, recommended practice for minimum weld size should be followed. Some transformers may have a false bottom or skids incorporated into the transformer case or have external stiffening members or mounting plates. If welds are made to these surfaces, local heating will generally not be a concern. Some transformers have base plates that are an inch thick and can accommodate welds, while others may have a relatively thin skin, with internal members to provide the needed structural support. In the latter case, welding to the base would be ill advised. If possible, the manufacturer should be contacted to get information on these issues. There is no consensus among utilities as to the appropriateness of welding directly to the transformer tank. Some utilities think that welding directly to the tank should not be done unless approved by the manufacturer. Other utilities think that welding can be done, but with caution. Large single-pass welds that generate high heat should be avoided. The transformer should be filled with oil, so that internal heat can be quickly carried away. In addition to anchorage, retrofit bracing of transformer radiators often entails welding brackets to the transformer case.

Figure 5-33 shows a retrofit of transformer anchorage. Retrofits are often less than optimal solutions and yet may still be adequate to address the problem. In the case at hand, it was difficult to get the drilling rig close to the transformer to drill anchor bolt holes. Therefore, an extension plate was welded to a mounting flange provided by the manufacturer. For this configuration, there is no potential problem of heating internal parts or concern about the ability of the case to carry seismic loads. When subjected to a horizontal ground motion, the anchorage on one side will experience a signifi-

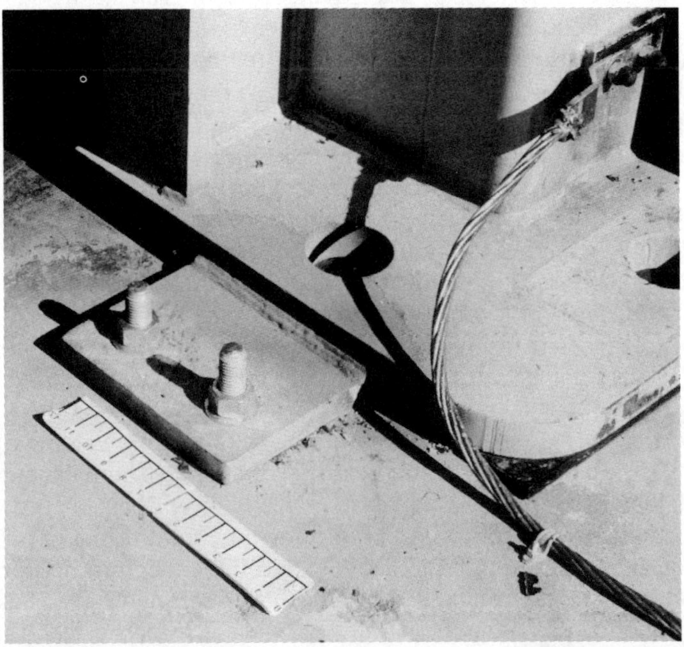

FIGURE 5-33. Retrofit of a Transformer Anchorage.

Note: A plate was welded to a mounting flange on the transformer and then bolted to the foundation slab.

cant vertical load resisting the overturning moment. The bolts are some distance way from the weld, so they are subjected to prying action, and bending the extension plate adds flexibility to the anchorage. Should the edge of the transformer lift off of the slab because of overturning moments and vertical ground motions, large impact forces could be generated when it drops back to the slab. Impact can generate large accelerations that could damage the porcelain bushings and lightning arresters.

There have been cases of expansion bolt failures. Because of potential problems with quality assurance and the sensitivity of expansion anchors to proper installation, they are not recommended in applications where large axial loads must be resisted. This recommendation does not apply to undercut expansion anchors.

Figure 5-34 shows another approach to a retrofit anchorage. In this case, an anchor bracket was welded to a stiffening flange on the transformer. Note that this plate has a gusset to stiffen the anchorage. One problem with this approach is the difference of stiffness along the weld between the bracket and the transformer flange.

Figure 5-35 is a schematic diagram that illustrates important design features. The figure shows two methods for anchoring a transformer to its foundation slab. Case A has several undesirable features. The anchor bolts are set back from the vertical part of the bracket increasing the flexibility of the anchorage and allowing the transformer to lift off of its slab. The setback also increases the prying action on the bolts to lifting loads. For loads tending to lift the transformer, the bolt pattern will cause the bolts closest to the transformer to carry most of the load. The weld will be subjected to bending loads along the axis of the weld. For horizontal loads, the vertical flange of the bracket gives a flexible anchorage. Case B shows an improved design.

FIGURE 5-34. A Bracket with a Gusset Was Welded to a Stiffener on the Transformer Case and Bolted to the Foundation Slab.

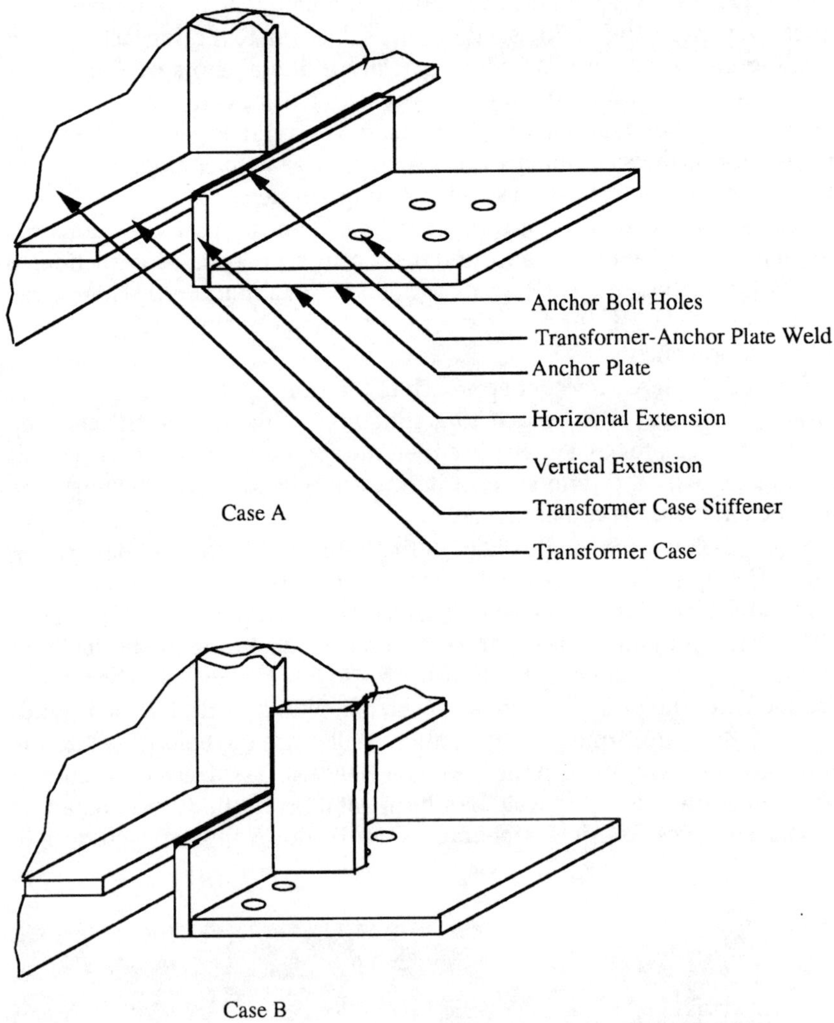

FIGURE 5-35. Schematic Diagram of Anchor Plate Designs To Illustrate Issues of Stiffness, Prying, and Stress Distribution along Welds.

Note: Weld sizes should conform to requirements of AISC or AWS.

The gusset stiffens the anchorage and resists bending along the horizontal weld. The bolts have been placed close to the transformer, reducing prying and distributing loads on the bolts more uniformly.

One concern with welding to the transformer case is knowing whether the case can withstand the concentrated load if there is an earthquake. In the Northridge earthquake, a hole was torn in a transformer case at an anchor weld. This problem can be eliminated by using the lifting lugs provided by the manufacturer as anchor points. These points protrude from the transformer case so that the problem of heating the transformer inner surfaces is eliminated, and the lifting lugs were designed to withstand large loads. Two examples of this type of retrofit are shown in Figures 5-36 and 5-37. These were added to transformers after the welds to embedments had failed or showed signs of distress. These retrofits were designed primarily to resist overturning moments and vertical loads. As can be seen in Figure 5-36, one of the disadvantages of using the lifting lugs is that their location may not be convenient for transferring the loads to the foundation slab.

The schematic diagram in Figure 5-38 illustrates design features of the system shown in Figure 5-37. Note that a mirror image of the items shown also exists on the back side of the lifting lug. The headed studs and base plate are fabricated and grouted into a pocket, and holes are cut in the concrete slab. The chair is welded to the lifting lug. The anchor bolt and the bolt–base plate tie are fabricated and welded together. When the grout has cured, the tie plate with attached anchor bolt is welded to the base plate, and the base plate is welded to the base of the transformer. The tie plate should be welded on both sides to maintain symmetry and to reduce bending loads.

FIGURE 5-36. Transformer Retrofit Anchorage Using a Transformer Lifting Lug.

FIGURE 5-37. Transformer Retrofit Anchorage Using a Transformer Lifting Lug.

The welds to the original embedments were repaired. Dynamic lifting loads are carried primarily by the anchor bolt and studs. Part of the shear load is carried by the new base plate and studs, and the remainder by the original steel embedments (with repaired welds).

All surfaces should be properly protected from corrosion. The holes for the headed studs should be roughened and thoroughly cleaned. A non-shrink grout must be used. While initial installation would be easier without recessing the base plate, recessing it is recommended. When the transformer is removed, the weld between the tie plate and base plate can be easily cut, leaving the foundation slab flat.

The installation described used an epoxy grout to secure the studded base plate to the foundation slab. Concrete grout or undercut expansion anchors are also options. The selection of the type of anchor must be made for many of the retrofit and new installation methods discussed in this docu-

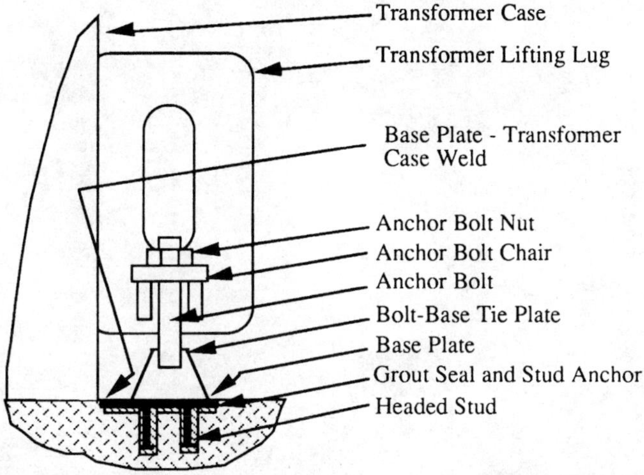

FIGURE 5-38. Schematic Diagram of Anchor Using a Transformer Lifting Lug.

ment, so factors that might influence the choice are discussed next. In this installation, the base plate conceals the grout and the holes when the base plate-headed stud assembly is inserted. Variations in mixing concrete grouts will influence their ability to flow. The uniformity of epoxy grout viscosity may yield a more reliable anchor when the action of the grout cannot be observed. The use of expansion anchors is another option; however, performance of some types of expansion anchors is highly dependent on the quality control of the installation. Undercut expansion anchors, while more expensive to install, have given more consistent results. Another concern about expansion anchors is that water may accumulate in the hole in outdoor installations, so there is a potential of anchor corrosion for both carbon and stainless steel. This should be considered in light of the fact that the presence of corrosion cannot be evaluated after installation.

A retrofit approach used by another utility is pictured in Figure 5-39, which shows one of four telescoping members that provide a load path from the transformer lifting boss to the foundation pad. The telescoping design can accommodate different transformer configurations and variations in the foundation slab. The two telescoping sections are welded together in the field after they are fitted to the transformer and slab.

5.7.2.3 Rail-Mounted Transformers. *5.7.2.3.1 Earthquake Performance.* One of the traditional methods for supporting transformers was to mount a transformer on a carriage with wheels and support the wheels on rails. This method of installation was used before means were developed for moving heavy equipment. A new installation of a rail-mounted transformer today would be unlikely. While this practice has not been used in California for

FIGURE 5-39. Field-Welded Telescoping Members Provide a Load Path from the Transformer Lifting Boss to the Foundation Pad.

some time and most existing units have been retrofitted, inadequately anchored, rail-mounted transformers are common in other parts of the country. The rails are often secured to an elevated platform so that the transformer could be rolled onto a transfer cart in front of the bank of transformers. In some installations, the rails supporting the transformer are at grade.

The wheels are restrained from rolling on the rails with chocks, which are typically bolted to the rail in front and behind the transformer. There are many examples where the chocks failed to restrain the transformer, and it fell from its supports. The chocks restraining some transformers are held by the clamping action and friction of a bolt, as indicated by a close-up view of a chock that was forced off the end of the rail (Figure 5-40). The transformers toppled from the raised pedestal, and this resulted in damaged bushings, radiators, and bus work (Figure 5-41). Note that the carriage on the first transformer remained attached to the transformer while the second and third transformers separated from their carriages. A positive attachment of the chocks to the rails would improve their performance. The chocks can be bolted through the rails or welded to the rails. The chock can fail, even when bolted, because of prying and shear action (Figure 5-42). While most rail-mounted transformers have some type of restraint, some restraints are woefully inadequate, as illustrated by the wooden wedge shown in Figure 5-43. In addition to being crushed, back and forth movement could dislodge the wedge, giving the transformer complete freedom to roll off the rails.

FIGURE 5-40. A Chock Bolted to the Rail Was Forced off the End of Rail, Allowing Transformers To Fall from Support.

Note: The chock was placed back on the pedestal before this picture was taken.

FIGURE 5-41. Rail-Mounted Transformers That Rolled off the End of the Pedestal and Tipped Over.

Note: The chocks were pushed off the end of the rail, allowing the transformers to fall, causing severe damage.

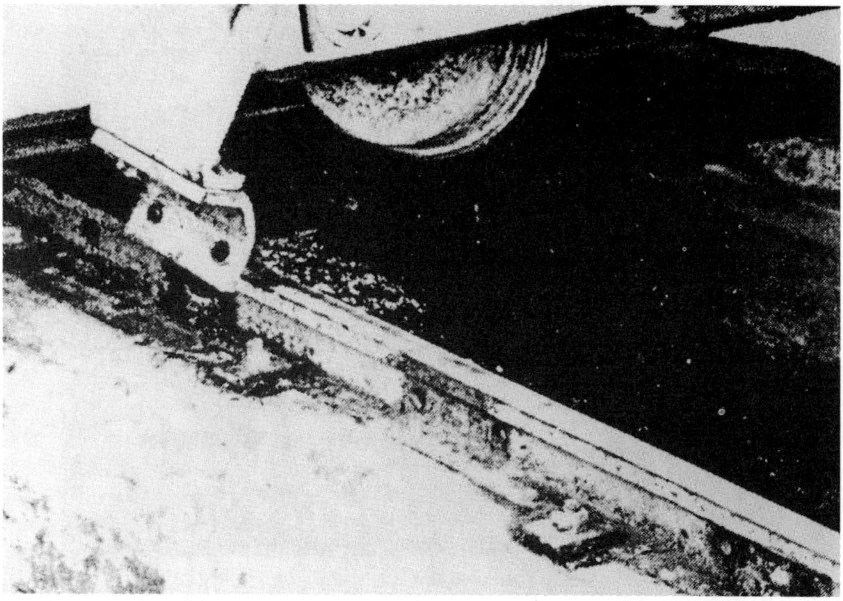

FIGURE 5-42. Chock Bolted to the Rail To Provide Positive Restraint.

Note: Prying, bending, and shearing action were sufficient, however, to cause the bolts holding the chock to fail.

FIGURE 5-43. Wooden Wedge Used as Chock To Restrain Transformer.

Some utilities use rails to support transformers in a different way. In this method, rails are placed about a foot apart over the transformer footprint, and the transformer simply rests on the rails. This allows the area under the transformer to dry out and provides increased air circulation around the transformer. Of course, this provides no anchorage.

A chock can prevent movement along the rails and still not provide lateral restraint. A system to limit the lateral loads that must be resisted by the rails and their anchors uses a key between the rails and the carriage under the transformer (Figures 5-44 and 5-45). Horizontal transverse loads are carried by the key and the pedestal, rather than the rails or their supports. There is a very small gap between the key and the wheel housing. The key has a large angle forming its upper corner to distribute contact loads. The massive wheel support has large gussets and is bolted to the carriage, which in turn is welded to the base of the transformer. The load path between the transformer and the slab is assured. Bolts of 2.5 cm (1 in.) secure the sides of the chock to the rails. A single 2.5-cm (1-in.) bolt secures the chock to its side supports. The chock is cut to fit the curvature of the wheel and extends about two-thirds of the way to the axle.

The large, well-anchored chocks and the key shown above do not provide positive anchorage to the transformer, which is still free to lift off the rails and tip over.

5.7.2.3.2 Retrofitting Rail-Mounted Transformers. There are several methods for providing positive anchorage to rail-mounted transformers. One method of marginal effectiveness is to weld the wheels to the rails. This pro-

FIGURE 5-44. Rail-Mounted Transformer Restrained by Chocks and Concrete Keys under the Carriage.

Note: The rails are mounted to a slab at grade.

FIGURE 5-45. Close-Up View of the Wheel Housing and Concrete Key under the Transformer Shown in Figure 5-44.

Note: The concrete key restrains the wheel-carriage assembly laterally.

vides a positive connection, and there is some vertical restraint. It is an easy and inexpensive fix that can serve where small seismic loads are expected or as a temporary fix until a more secure method of anchorage can be implemented. The problem with this method is that the load path between the transformer and the slab was not engineered. The transformer must be well anchored to its carriage but,as shown in Figure 5-41, this is often a poor connection using tack welds. Also, the weld length between the wheels and the rails is short, so that its capacity may not be adequate. It may also be difficult to generate enough heat to get a good weld.

Longer welds can be obtained by welding a plate to the wheel and to the rail. However, there have been difficulties in welding to the rails, and the carob content of the steel should be checked to ensure that a sound weld can be made.

The loads transferred to the rails must pass through the rail anchors to the pedestal. There are cases where rail supports are adequate, as shown in Figure 5-46. Each of these relatively small transformers has four bolts holding the transformer carriage to the two brackets. The brackets are anchored to the rail, and eight bolts hold the rail to the transformer pedestal. Ample welds that cannot be seen in the picture secure the transformer and carriage. The width of the pedestal provides adequate edge distance for the rail anchor bolts. Rail anchors often use friction clips, however, and have inade-

FIGURE 5-46. Transformers Can Be Anchored through Their Support Rails If the System Is Properly Engineered.

quate edge distance. Figure 5-47 shows the original rail anchorage and the retrofitted anchorage. Note that the anchorage could have been improved by welding the friction clips to the rail. In the current situation, the load directed toward the friction clip must be carried entirely by the clip shown, and loads are put on the bolts with marginal edge distance. If the clips were welded to the rail, the clip on the other side of the second rail would also carry load, and the edge distance would be across the remainder of the pedestal width. In general, while the total capacity of the rail anchorage may be adequate, the loads are concentrated at two locations. The retrofit addresses this issue. The use of friction clips is discussed further in Section 5.12, Circuit Breakers.

The retrofit of pedestal-supported, rail-mounted transformers should address several issues. The transformer should be restrained in the longitudinal, transverse, and vertical directions. Loads should be transferred to the foundation. The entire load path should be engineered. It is desirable to avoid large axial loads on expansion or grouted-in-place bolts. Figure 5-48 illustrates one approach to retrofitting. Four diagonal braces are welded to the transformer undercarriage and anchored to the pedestal with bolts through the pedestal. The brace, cut from appropriately sized tubular steel, will carry longitudinal, transverse, and lifting loads, including overturning moments. The rails will support the transformer weight and carry part of the

FIGURE 5-47. A Transformer Anchored to Its Support Rails by Improving the Rail Anchorage System.

transverse load. The use of bolts through the pedestal eliminates the concern about axial loads on expansion anchors and the issue of installation quality assurance. It is vital that the weld between the transformer case and the carriage is adequate. Because of the large contact surface between the carriage and the case, a small, long weld can be used. The recommended practice for minimum weld size should be followed as specified by the American Institute of Steel Construction (AISC) and the American Welding Society (AWS). This distributes the load on the transformer case and reduces the heat introduced into a concentrated area of the case; it also will generally be located near the transformer base-wall boundary, so the load path should be relatively stiff. The fabrication of the brace is relatively simple. In some transformer configurations, the brace may be welded directly to the transformer case, so concerns about the load path through the carriage can be eliminated. Clearly, the pedestal must be secured to the foundation slab or of such a design that it is structurally secure. For very heavy transformers, very large braces may be required to restrain lateral loads. In this case, multiple braces could be used or diagonal braces could be added under the transformer.

Another approach to retrofitting a rail-mounted transformer is shown in Figures 5.49 and 5.50, where there is no pedestal. A similar approach could be used for a pedestal-mounted system and has the advantage that loads are transferred directly to the transformer foundation slab. The bracing system at one end shown in Figure 5-49 supports loads in all three directions, while

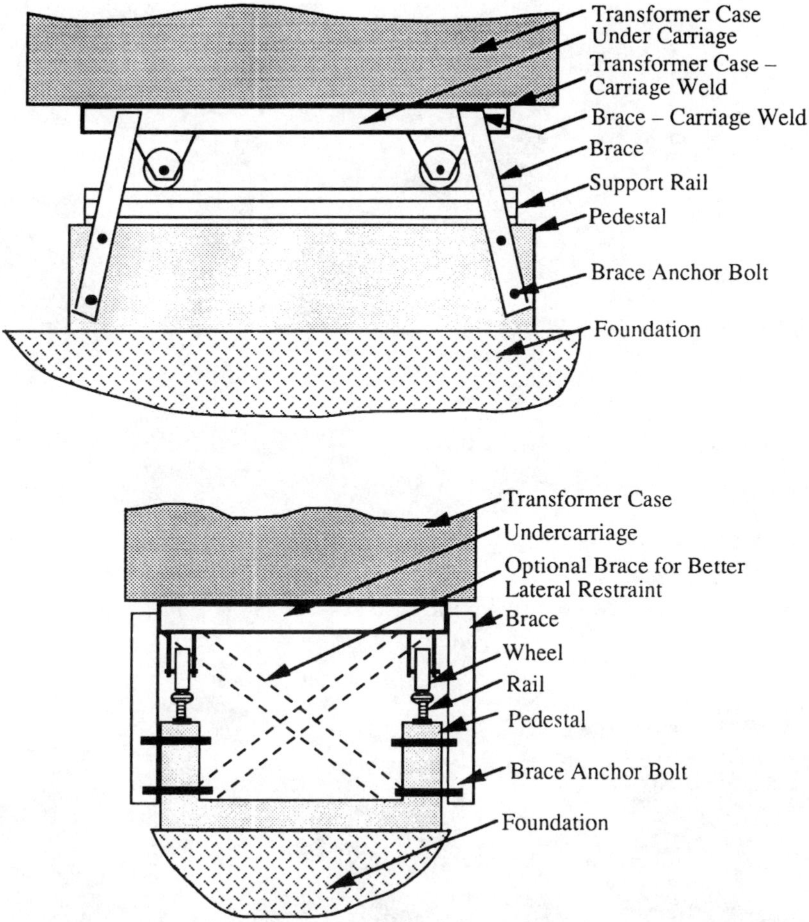

FIGURE 5-48. Schematic Diagram for Retrofitting a Pedestal-Supported, Rail-Mounted Transformer Illustrates Key Features of the Design.

that shown in Figure 5-50 at the other end is not designed to take loads parallel to the rails. The weight of the transformer is carried by the rails. The braces are welded directly to the transformer case. Undercut expansion anchors are used to anchor the framing members to the slab.

5.7.2.4 Anchored, Slab-Mounted Transformers. Unanchored, slab-mounted transformers and methods for their retrofit were discussed above. Many methods for anchoring slab-mounted transformers have been used, and their performance will now be reviewed. The methods for retrofitting unanchored transformers would also be applicable to the installations discussed below.

FIGURE 5-49. This Frame Transfers All Loads from the Transformer to the Foundation Slab.

FIGURE 5-50. This Frame at the Other End of the Transformer Shown in Figure 5-49 Is Designed To Resist Transverse and Vertical Loads Only.

5.7.2.4.1 Earthquake Performance of Anchored, Slab-Mounted Transformers. Several methods used to anchor slab-mounted transformers have failed. The review of these failures can provide insight into features to be avoided in new designs.

The failure of bolts used to anchor transformers has been very common. Figure 5-51 shows one of four restraints used to anchor a transformer. The

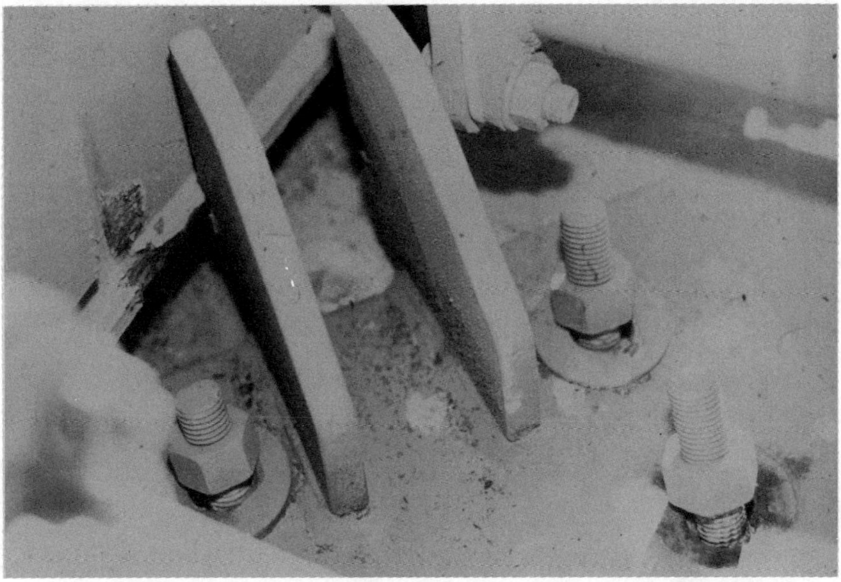

FIGURE 5-51. Failure of Bolts Used to Anchor Transformers.

Note: The four bolts anchoring the restraint sheared off. Since the transformer was not positively attached to the restraint, only restraints on one side of the transformer carried loads. Gashes in the transformer bottom plate were made when the transformer impacted the restraint.

four restraints, two on each side of the transformer, have a small notch to match the profile of the base of the transformer. Each restraint had four anchor bolts. The peak ground acceleration at the site exceeded the design level; however, the energy in the motions was moderate. Small changes in the design may have improved performance. The restraints were not welded to the transformer case to address the concerns noted above. As a result, loads directed toward the restraint had to be resisted by the restraint on that side of the transformer. Welding the restraint would have allowed restraints on both sides of the transformer to share the load. When the restraints on one side are lost, the transformer is free to slide. As the transformer slides back and forth during the earthquake, impact loads generated when contact is made with the restraint can damage bushings. In this case, the transformer bushing leaked, but the transformer remained in service. Figure 5-52 shows the restraint on the other side of the transformer, which has moved away from the restraint. Near the restraint in the figure is an oil drain pipe and valve. Had the transformer moved a few centimeters longitudinally, the drain pipe could have been sheared off, allowing the transformer to drain. This would have taken the unit out of service and required a major oil clean-up. At the top of the picture, the other restraint

FIGURE 5-52. Restraint Bolts on the Other Side of the Transformer Shown in Figure 5-51.

Note: The restraint bolts on the other side of the transformer sheared, allowing the transformer to slide. The oil drain pipe near the restraint could have been damaged if the transformer had moved a few centimeters in that direction.

can be seen; the weld between the hook and the base plate has been severely torn.

In the 1995 Kobe, Japan, earthquake, several transformers at different substations sheared their anchor bolts. The transformer in Figure 5-53 sheared the eight 4.5-cm (1.75–in.) bolts that secured it and slid more than 30 cm (1 ft). Control cables for this transformer enter through a hole in the foundation slab. The movement of the transformer almost sheared off the control cables (Figure 5-54). A small reactor connected to the low-voltage output also sheared its bolts, but slid in the opposite direction. The total relative movement was more than 60 cm (2 ft), and the lack of this magnitude of slack in the connection caused the low-voltage bushings to fail.

The performance of transformer anchor bolts has been rather poor. While the cause is not fully understood, one speculation is that the clearance between the bolt and the hole in the transformer anchor bracket for the bolt allow the transformer to have a moderate velocity when the transformer impacts the bolt. The geometry of the connections puts the bolt into shear with very little flexibility. Thus, the bolts abruptly stop the transformer, generating impact forces in the bolts and the entire transformer much larger than would be expected from the ground accelerations.

FIGURE 5-53. Eight Bolts Holding This Transformer Sheared, Allowing the Transformer To Slide More Than a Foot.

FIGURE 5-54. Control Cables Entering the Transformer through the Foundation Slab Were Almost Sheared Off Because of the Movement of the Transformer.

Figure 5-55 shows a transformer at the same site as that shown in Figure 5-51 with clips welded to slab embedments. The overturning moments caused the transformer to tip, placing lifting loads on the clips. The load was sufficient to deform the embedded plate and start to tear the clip–plate weld. One disadvantage of this configuration is that the weld will tend to "unzip". While this dissipates energy, it can lead to catastrophic failure of the anchorage. Also, when the transformer drops back to the slab, large impact loads will be generated.

The most common current practice for anchoring transformers is to weld the base of the transformer to a steel embedment in the foundation slab. Three failure modes of this system have been observed, although it has generally performed well. One failure has been due to poor quality or inadequate strength welds, which allowed one end of the transformer to move 35 cm (14 in.) (Figure 5-56). There were two interesting observations associated with this failure. While the welds on each side at one end of the transformer failed and the transformer moved, the welds at the other end of the transformer did not fail. The transformer case was fabricated from heavy plate, and its box configuration gives the impression that the case would be very stiff. On the side of the transformer opposite that shown in Figure 5-56, a mop head was found under the transformer in a position suggesting that

FIGURE 5-55. Transformer with Clips Welded to Slab Embedments.

Note: Clips welded to embedments were severely loaded because large overturning moments and the tipping of the transformer. The weld to the embedment was partially torn and embedment plates were deformed, but the transformer was restrained. The lifting action probably caused impact loads when the transformer fell back to the slab.

FIGURE 5-56. Anchor Welds Failed on One End of This Transformer, and It Moved 35 cm (14 in.); Anchor Welds at the Other End Did Not Fail.

the bottom of the transformer had lifted about an 2.5 cm (1 in.). This indicates that one end of the transformer had moved laterally 14 in. and lifted about an inch, while the welds at the other end stayed intact.

There have also been cases where there were signs of distress in the embedment, but none have failed. Figure 5-57 shows cracking and spalling of concrete and grout around a steel embedment used to anchor a transformer. This damage was caused by ground motions with a peak acceleration of only 0.15*g*. A retrofit to this anchorage is shown in Figure 5-37. Speculation on the causes for distress in the embedments includes theories that the thickness and stiffness of the plate were inadequate, the points anchoring the plate in the concrete were too far apart, and the anchor plate with its embedments did not extend far enough on each side of the weld connecting the transformer to the plate. If the transformer is welded near the end of the embedment, lifting loads will be borne primarily by the last set of headed studs.

Finally, one transformer, which had a retrofitted anchorage with the case welded to an embedment, ripped the transformer case, causing an oil leak.

An integral part of the anchorage of any transformer is its foundation slab. Tilting of a slab was illustrated in Figure 5-8. However, even when no permanent tilting results, there have been indications of the rocking of foundation slabs. Low-voltage, rigid-bus connections and other conductor connections on several transformers were severely damaged, and scrape marks on sidewalks adjacent to the transformers suggest that there was soil-structure interaction of the massive transformers and their foundation slabs with the surrounding soil. Some of these effects were observed at sites where there are no indications of soil liquefaction. Relatively small rotations can cause increased

FIGURE 5-57. Concrete and Grout around the Embedment Cracked When Peak Ground Acceleration Was Only 0.15g.

inertial loads on bushings and significant movement at bushing connections, which can increase conductor interaction loads on the bushings.

5.7.2.5 Recommended Practice for New Installations. The recommendation for new transformer installations is to weld the transformer to embedments in the foundation slab. This involves three topics: foundation design, embedment design, and welding to the embedment.

5.7.2.5.1 Foundation Design. The foundation design for a transformer should address the following topics.

- The foundation slab must be thick enough for the embedments to develop the needed pull-out strength.
- The foundation slab should be dimensioned to withstand overturning moments, or else other means to improve stability should be considered.
- The foundation slab must be strong enough to withstand bending moments for that part of the slab that extends outside the transformer footprint.
- The area of the foundation slab or the use of piles must be such that the bearing capacity of the soil is not exceeded. Where appropriate, the soil under the transformer should be compacted prior to pouring the foundation, and poor soils may need to be replaced with engineered fill.

- Where conditions for soil-structure interaction exist, various options to reduce potential impacts should be considered. These include providing generous slack to conductor connections; using seismically robust bushings; and using piers, piles, or other means to control movements.

Many of the foundation slabs that were designed to standards much less stringent than those used today have survived significant earthquakes without damage. Most of the problems of differential settlement and the effects of soil-structure interaction can be addressed by providing adequate slack to conductors connected to the transformer. The number of sites that are vulnerable to soil liquefaction and that have been subjected to strong ground motions is very limited. Thus, the need to more thoroughly address soil liquefaction and lateral spreading beyond providing adequate slack to conductor connections remains uncertain at this time.

5.7.2.5.2 Embedments. An embedment design should satisfy several criteria. In determining anchorage loads, friction between the transformer and the slab should be neglected. Lateral loads as well as overturning moments should be considered. While transformers tend to be symmetrical, the position of radiators and the placement of anchors may introduce torsion loads that should be accounted for in the design. The embedment must be stiff enough to transfer loads to points where the embedment is anchored to the concrete and to resist tearing and bending stresses. The embedded surface that is to be welded to the transformer must be wide enough that the size of the weld to carry the design loads is consistent with the strength of the members that are being welded. The longitudinal dimension of the embedment is usually perpendicular to the edge of the transformer that is to be welded. The embedment should extend well beyond each side of the weld. It is advisable to extend the embedment well outside of the transformer footprint, so a large replacement transformer could use the same slab and embedment.

The location of the embedment must be coordinated with the transformer manufacturer to assure that the transformer has been designed to withstand concentrated loads at the anchor points. Stiffening plates, brackets, or other features might be added by the manufacturer to facilitate field welding. It would be desirable for the industry to standardize the placement of transformer anchor points so that the location of embedments would be consistent among transformer models and manufacturers.

Three embedment configurations have been used, and other alternatives can be used that meet the design criteria. Embedments that have been used include plates with headed studs, inverted channels with headed studs or welded rebar wings, and wide-flange beams with holes for rebar to pass through. There have been problems with deformation of plates due to the heat generated when the concrete is curing; thus, thick plates or plates with a stiffener are suggested. It is recommended that studs extend below the slab rebar. Tests indicate that headed studs perform much better than "J" or "L" bolts. The American Concrete Institute recommends the use of headed studs only . While the embedment can be welded to slab rebar, this is seldom done. When the concrete is poured, it is important that voids under the embedment be eliminated. In some cases, when inverted channels are used, holes are cut in the web to avoid leaving voids underneath. Conventions for bolt separation and edge clearance should be followed. The embedment should be flush with the surface of the slab and never below it. The embedment should be given appropriate corrosion protection treatment.

5.7.2.5.3 Welding to Embedments. Typically, the edge of the transformer case is welded to the embedment. Situations arise where an extension plate or angle is used to transfer the load from the transformer to the embedment, as illustrated in Figure 5-58. The stiffest anchorage and preferred method of attachment is to weld the transformer directly to the embedment. The advantage to this method is that the transformer is rigidly held to the foundation slab so that lift-off and subsequent impact is avoided. If extension plates are needed, they should be stiffened as illustrated in Figures 5.35 and 5.58. If overturning moments lift the transformer, the fillet weld would also be subjected to bending along the axis of the weld, which is an unacceptable condition. If an unstiffened plate is welded along the edges perpendicular to the edge of the transformer as a means to increase the stiffness of the anchorage, those welds will tend to unzip when subjected to lifting loads. All welds should be carefully cleaned and given corrosion protection treatment. One of the difficulties frequently encountered when welding the base plates of transformers to embedment plates in the foundation slab is that gaps between the plates may occur that are due to manufacturing or construction tolerances. Gaps should be shimmed, and weld details should specify adequate weld penetration.

Figure 5-59 illustrates important design considerations for transformer anchorage. Not covered in the illustration are issues related to soil bearing capacity, overturning moments, slab strength, soil liquefaction, and soil-structure interaction.

5.7.3 Bushings

Bushings are usually mounted on a turret, that is, a small cylindrical pedestal, which is mounted on top of the transformer case or mounted off the side of the transformer. The bushings are often mounted with the longitudinal axis at an angle of about 15° from vertical. In this case, there is a static moment applied to the base of the bushing. A bushing is constructed with a conductor surrounded by an insulating material, usually porcelain. A mounting flange divides the insulating material with the lower part con-

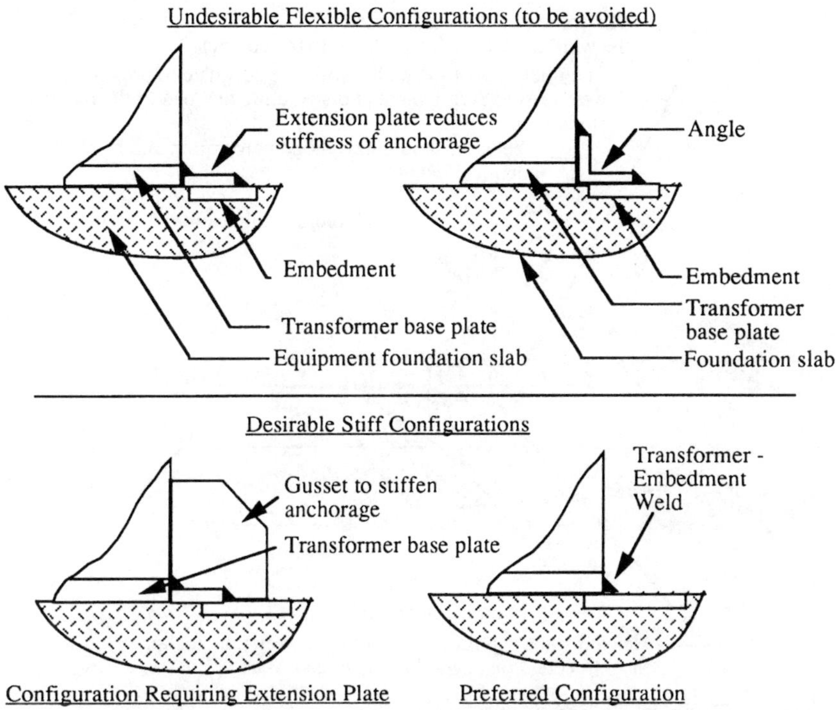

FIGURE 5-58. Methods of Welding a Transformer to an Embedment.

Note: The use of extension plates or angles to weld a transformer to an embedment should be avoided as they add flexibility to the anchorage. Also, the headed studs on embedments are not shown.

tained in the transformer case and the upper part extending above the turret. The end of the conductor that extends beyond the upper porcelain is a binding post for making external connections. Most of the earthquake problems with bushings involve the upper porcelain.

Porcelain bushings have several mechanical designs that can influence their earthquake performance. The most common design uses post tensioning to hold the upper porcelain member to the metal mounting flange; a gasket forms a seal between the porcelain and the flange. A second, similar design has the porcelain seated in a shallow pocket in the metal flange. A third type has the porcelain cemented into a pocket in the flange. Some bushings are fabricated from two or more sections, with the sections separated by gaskets and held together by the same post-tension forces that hold the porcelain to the flange. New high-voltage designs have included fiberglass-reinforced polymer (FRP) and polymer concrete bushings.

In addition to the bushings for the primary and secondary voltages, there may be bushings for tertiary voltages.

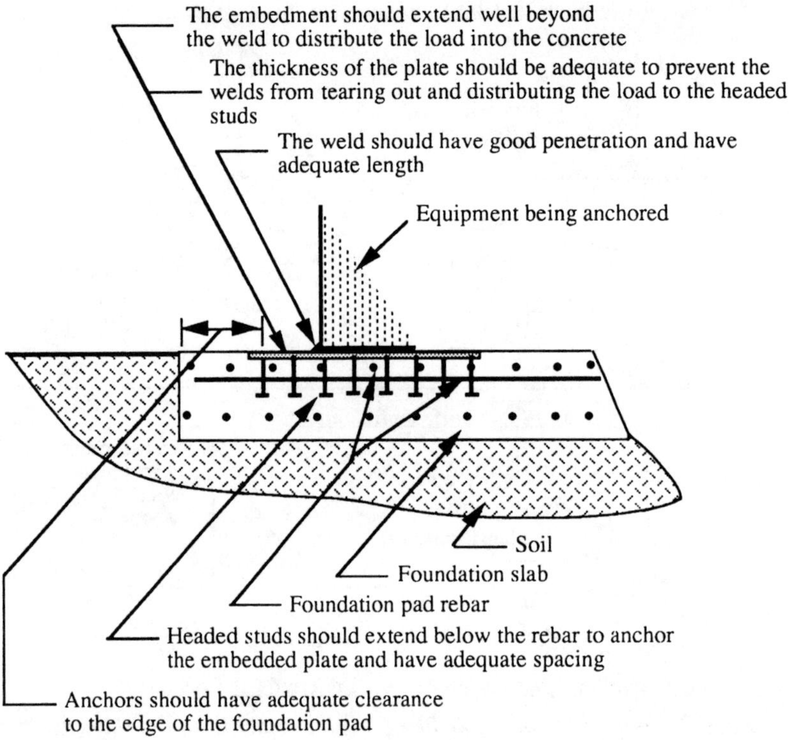

FIGURE 5-59. Important Features in Designing a Transformer Anchorage.

5.7.3.1 Earthquake Performance of Transformer Bushings. Inertial loads on bushings are due to ground vibrations, which may be amplified by the dynamic response of the transformer, the turret supporting the bushing, and the rocking response of the transformer associated with soil-structure interaction. There can also be interaction loads from the conductors. Bushings may also be influenced by the effect of a failed lightning arrester. Issues related to lightning arresters are discussed in Section 5.9.

Several types of bushing failures have been observed. Most failures are associated with bushings that were held to a flat flange under a post-tension load. The most common failure is due to the upper porcelain of the bushing moving relative to its support flange. There can be an offset of the porcelain relative to its flange (Figure 5-60). Typically, offset porcelain leaks oil. In some cases, the gasket between the bushing and the flange protrudes (Figure 5-61). There have been many cases where there were signs of fresh oil stains, which indicated leaks had occurred during the earthquake, even though they had stopped by the time of the postearthquake inspection. For this type of bushing, there can be significant variations in earthquake performance among bushings from different manufacturers. Bushing maintenance prac-

FIGURE 5-60. Porcelain of a Bushing That Slipped Relative to Its Mounting Flange, Typically Resulting in Oil Leaking from the Bushing.

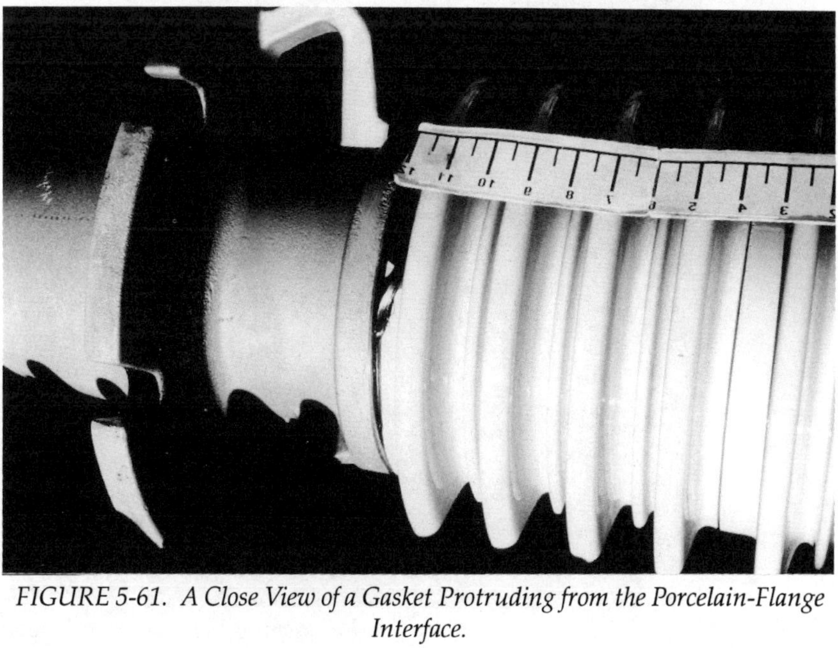

FIGURE 5-61. A Close View of a Gasket Protruding from the Porcelain-Flange Interface.

tices, primarily the post-tension load used in assembling the bushing, may also influence bushing performance.

Some bushings have sustained catastrophic failures, as shown in Figure 5-62. In some cases, it has been speculated that these failures may have been initiated when the upper porcelain of the bushing came in contact with the metal flange after the gasket had been blown out. In many of the failures, the design of the bushing has not been documented. The bushing shown in Figure 5-63 replaced one that had cracked longitudinally over the entire length of the bushing. The transformer was off line, and the presence of the crack only became apparent when dust stuck to a thin oil film that seeped out of the hair-line crack.

When a lightning arrester with a conductor connected directly to a bushing fails and falls, the imposed load placed on the bushing through the con-

FIGURE 5-62. Porcelain on a 500-kV Bushing Shattered.

ductor can damage the post at the top of the bushing. Several mitigation methods relating to lightning arresters are discussed in Section 5.9.

An unusual failure in a transformer bushing was the cracking of the metal flange that supported the bushing (Figure 5-64).

Three transformer explosions and fires have been attributed to earthquakes. Within a few weeks or months after an earthquake, a bushing exploded and set the oil in the transformer on fire, resulting in the total destruction of the transformer. The details of the failure mechanism are not understood; it may be that the failures are not related to the earthquake. A detailed evaluation of the bushings indicated that an arc had penetrated the paper insulation just below the mounting flange between the porcelain and the center conductor. There was no indication that the porcelain of the bushing had moved as a result of the earthquake.

FIGURE 5-63. This Bushing Replaced One That Had a Hairline Crack Running the Length of the Bushing.

FIGURE 5-64. The Metal Flange Supporting This Transformer Has Cracked.

High-voltage, fiberglass-reinforced, polymer bushings have recently been used to replace porcelain bushings. These bushings have been subjected to very severe vibration-table seismic testing and have performed very well. This material has also been used to replace other porcelain components, such as post insulators, and their response to limited seismic exposure has been good. Their cost is roughly comparable to porcelain bushings.

5.7.3.2 Mitigation and Retrofitting Transformer Bushings. Several approaches have been suggested for improving the earthquake performance of transformer bushings. First, it is interesting to note that no circuit breaker bushings have been damaged in the United States. The bushings on 230-kV bulk oil circuit breakers should be physically interchangeable with those on transformers if they conform to standards. Some of these bushings are constructed differently than transformer bushings because of the large dynamic loads they can experience under short circuit conditions and during circuit-breaker operations. While circuit breaker bushings cost more than transformer bushings, the cost differential relative to the cost of a transformer is small.

Providing adequate slack to conductor connections may reduce interactions loads and improve performance.

High-voltage bushings are often shipped by the manufacturer with a clamp around the bushing at the flange-porcelain interface to prevent slippage. At one facility, this clamp was inadvertently left on the bushing after installation, and this bushing did not fail while others at the site did. While this was not the only bushing at the site to survive, it suggests that the clamp may have improved the bushing's earthquake performance. This is an unverified method to improve the earthquake performance of bushings by preventing slipping and possibly keeping the gasket in position. This method may also increase seismic vulnerability, because it may induce a large contact load between the clamps and the porcelain, causing porcelain failure. Corrosion of the clamp may also increase pressure and cause porcelain failure.

It has been suggested that bushing performance may be improved by bonding the gasket to one of the surfaces of the flange-bushing interface. This may help keep the gasket in position and prevent it from being squeezed or blown out. Again, this mitigation method is untested.

Vulnerable porcelain bushings could be replaced with seismically more robust porcelain or fiberglass-reinforced polymer bushings. The seismic robustness of bushings can be determined through seismic qualification testing described in IEEE Standard 693.

5.7.4 Radiators

The radiators on transformers are used to cool the oil that circulates around the transformer coils. The oil circulation can be by natural convection. If circulation is by pumps, the transformer is referred to as a *forced-oil* type. If the radiators have fans, the transformer is referred to as a *forced-air* type. For the purposes of this document, four general classes of radiators are considered.

The first class is referred to as *manifold radiators*. Each radiator element is connected to a manifold, and the manifold makes the fluid and structural connection to the transformer case. This is illustrated in Figure 5-65.

The second class of radiator has each radiator element connected directly to the transformer, and it is referred to as a *nonmanifold radiator* (Figure 5-65). The upper and lower pipe penetrations for each radiator element support the element.

The third class of radiator is structurally independent of the transformer and is referred to as a *self-supporting radiator*. These radiators are typically associated with transformers that are surrounded by sound-insulating pan-

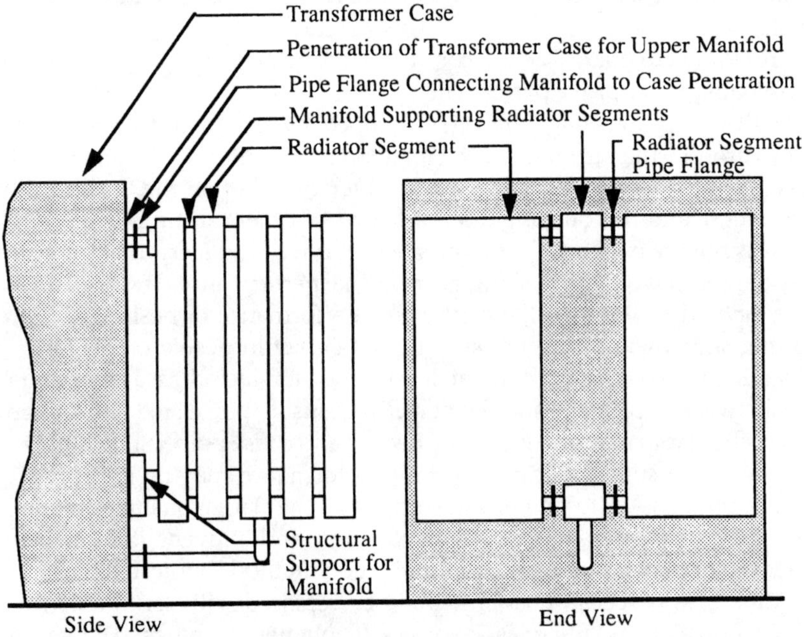

Manifold-Radiator Type Transformer

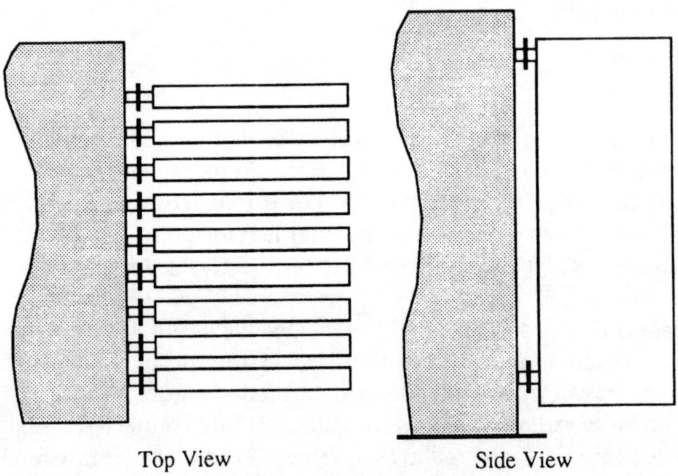

Non-Manifold Radiator Type Transformer

FIGURE 5-65. Illustrations of Manifold and Nonmanifold Type Radiators.

els to reduce transformer noise in nearby residential or commercial areas. The radiators are supported on their own foundation slabs outside of the acoustical enclosure (Figure 5-66).

The fourth class of radiator is distinguished by its small size; the volume of the radiator is less than 100 ft^3. These will be referred to as *small radiators* independent of the details of their design.

5.7.4.1 Earthquake Performance of Radiators. While radiator oil leaks are not unusual, most stop after the earthquake or can be stopped by tightening pipe flange bolts. Severe leaks have required that the transformer be taken out of service. Exposed oil from leaks presents a fire hazard, and spills require costly cleanup.

The manifold radiator is the most vulnerable. Typically, the lower manifold has a structural support that is separate from the piping penetration through the transformer case. The lower piping connection, which does not carry any structural load, is flexible relative to the structural connection. For the upper manifold, the pipe that serves as the penetration also serves as the structural support. While some manifold radiators have diagonal braces for the lower manifold, few have braces for the top manifold, unless they were designed for seismic loads. Oil leaks have occurred at flange connections at upper pipe connections. A major leak can require a costly and time-consuming cleanup (Figure 5-67).

FIGURE 5-66. Self-Supporting Radiators Are Positioned Adjacent to the Acoustical Enclosure That Surrounds the Transformer.

An indication of the magnitude of the vertical loads on a manifold is indicated by the deformation of the manifold support in Figure 5-68.

The performance of nonmanifold radiators has generally been good; however, in the Northridge earthquake, such radiators developed cracks in the upper pipe penetrations (Figure 5-69). The large radiator elements were

FIGURE 5-67. Oil-Contaminated Soil around the Transformer Had To Be Removed as Part of the Cleanup after a Severe Transformer Leak.

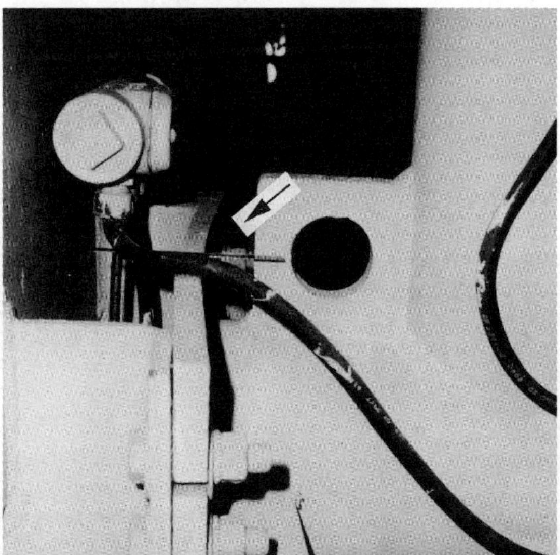

FIGURE 5-68. The Vertical Load on the Radiator Manifold Caused Its Support To Deform.

not braced on these transformers. Surprisingly, the ground motion at the site with the damage had a peak ground acceleration of only 0.15*g*.

Self-supporting radiators often use Dresser couplings on the pipes that connect the radiator to the transformer. Relative movement between the transformer and the radiator have caused leaks at the coupling. Figure 5-70 shows the pipe inside the acoustical enclosure with the Dresser coupling on

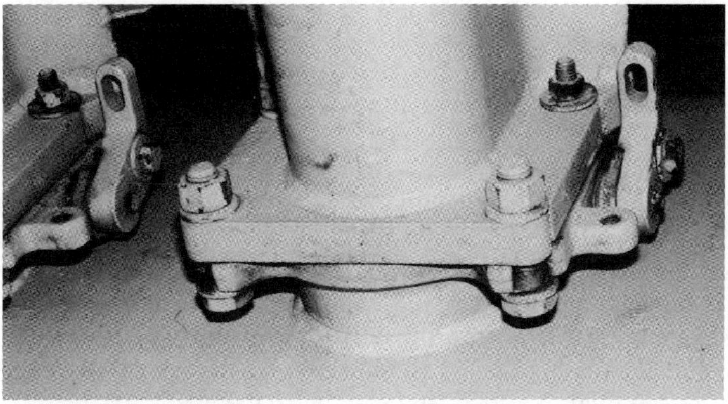

FIGURE 5-69. Close-Up View of the Lower Connection of the Radiator to the Transformer Body is Similar to the Upper Connection That Was Damaged.

FIGURE 5-70. The Dresser Coupling Connecting the Transformer and the Radiator Pipe Developed a Leak Due to the Relative Movement between the Two Units.

the lower pipe connection. Here, the upper coupling developed a leak and has been temporarily removed.

The performance of small radiators has been good. There may be small radiators of a vulnerable design that perform poorly, but they have not been observed in California or foreign earthquakes.

5.7.4.2 Mitigation and Retrofitting Transformer Radiators. The earthquake performance of radiators can be improved by reducing the structural loads that piping must carry. For manifold radiators, braces to the case to carry the radiator weight and reduce moments from lateral loads should improve performance. Figure 5-71 shows important features to improve earthquake performance.

It is important that the diagonal braces be very stiff because the pipe connections to the transformer are very stiff and will still carry most of the load if the brace is flexible. An example of a brace that provides marginal benefits is illustrated in Figure 5-72.

It is relatively easy to retrofit most nonmanifold radiators. There is a need for a structural member tying the radiator elements together. These members are often already installed on radiators without seismic designs. Diagonal braces can then be added between such a member and the transformer case. It is important that the brace connections are stiff and that braces are used at the top and bottom of the radiator elements. A retrofitted radiator is shown in Figure 5-73. Issues raised about welding directly to transformer cases in Section 5.7.2.2.2 also apply here.

For self-supported radiators, the pipe connections between the radiator and transformer should have adequate flexibility. The radiator anchorage should be stiff to reduce the demands for flexibility in the radiator-transformer connections.

5.7.5 Conservators

Conservators are large tanks that hold as much as 1,000 gallons of oil and are supported above the transformer case. They serve as a reservoir to accommodate the expansion of oil in the transformer case as it gets hot.

5.7.5.1 Earthquake Performance of Conservators. Two types of failure have been observed: the structural support system failed or the pipe connection between the conservator and transformer tank failed. One of the supports shown in Figure 5-74 failed; however, the tank did not fall and no oil leaked. In Chile, the bolts that held the frame supporting the conservator failed, and the tank would have fallen to the ground if it had not been supported by a firewall adjacent to the transformer.

There has been a failure of the pipe connecting the conservator to the transformer case resulting from the relative movement between the connec-

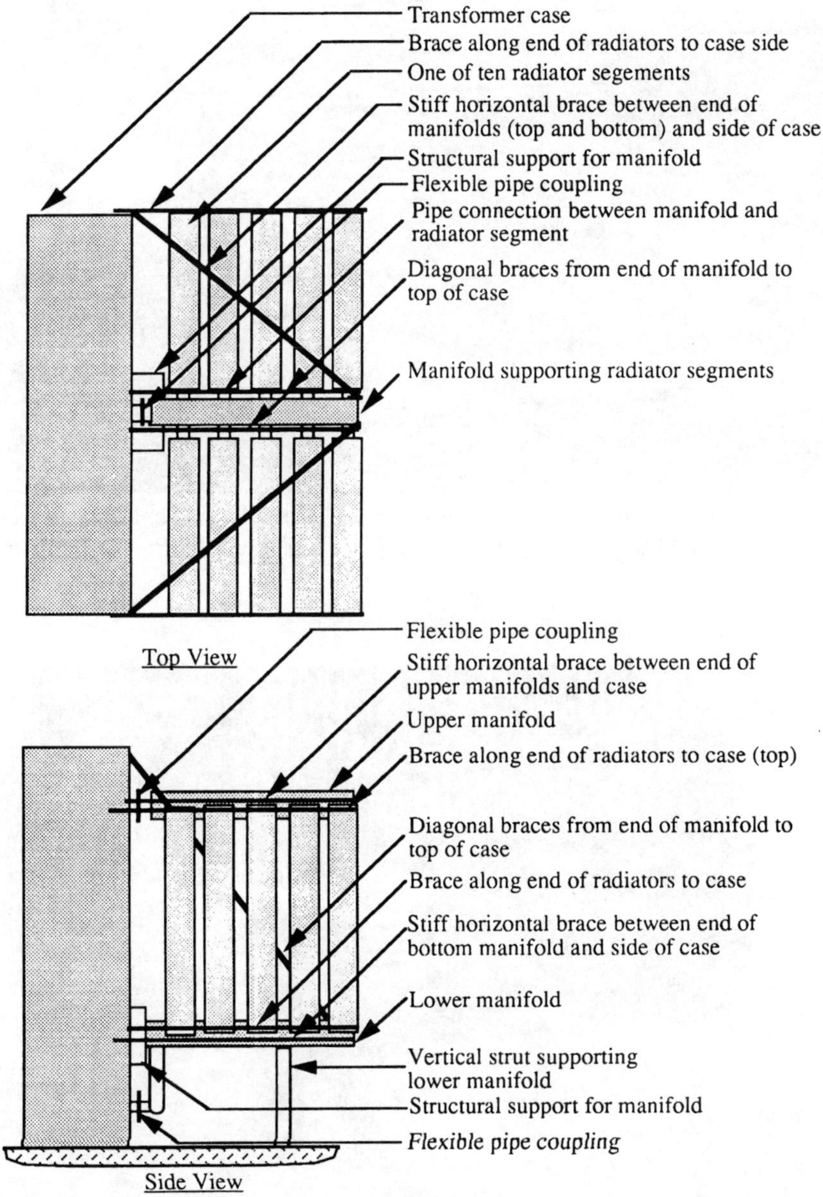

FIGURE 5-71. Schematic Diagram Identifying Important Features To Improve Radiator Earthquake Performance.

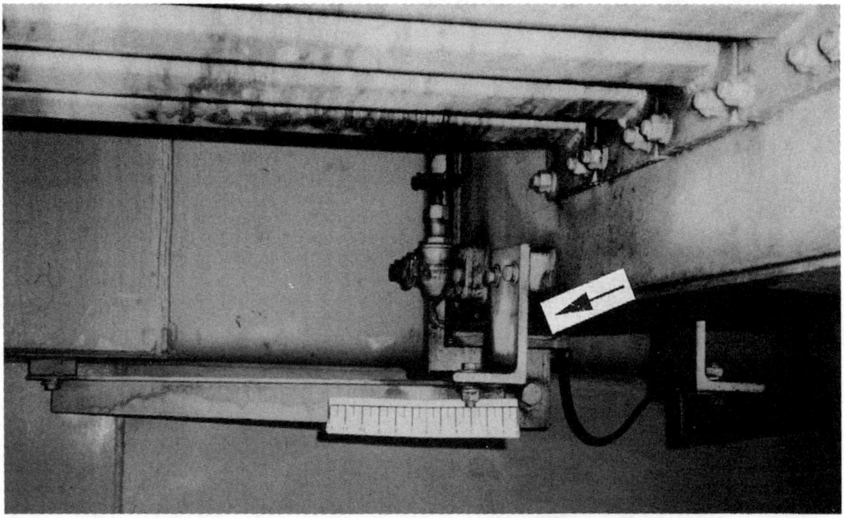

FIGURE 5-72. The Diagonal Brace between the Lower Manifold and Transformer Case Provided Marginal Benefits Because of the Flexibility of the End Connection.

FIGURE 5-73. Diagonal Braces Have Been Added to This Radiator To Reduce Moments Applied to the Pipe Connections.

FIGURE 5-74. The Front Support of the Conservator Failed, but Was Repaired prior to the Picture Being Taken.

tions. Several factors contribute to conservator failures. Some conservators are bolted to the transformer body with slotted connections. The seismic loads are such that the conservator moves relative to the transformer case. The second cause of movement between the conservator and the transformer case is the flexibility of the conservator to vibrations parallel to the axis of the conservator tank. For most designs, the small relative movement between the conservator and the transformer case does not cause a problem, but if the piping is stiff and cannot accommodate the movement, failure can occur. A piping failure resulted in an oil spill of about 1,000 gallons.

5.7.5.2 Mitigation and Retrofit Methods for Conservators. The failure of conservators has been relatively rare; however the consequences can be significant. An oil leak can require that the transformer be taken out of service, an expensive cleanup may be required, and there is a serious fire hazard should the oil ignite. For the few transformers that have vulnerable conservator-transformer piping and flexible conservator supports, a simple piping modification would eliminate the potential problem. If the conservator has to be drained for some other reason, modifying the piping would be simple and inexpensive.

5.7.5.3 Recommended Practices for Conservators. Clearly, the design of the conservator support and piping is the responsibility of the transformer manufacturer. The equipment seismic qualification standards should greatly reduce structural failures. However, the causes of leaks have been from relative movements between the conservator and the transformer, and these could still occur without structural damage. It is important that the piping connection between the conservator and the transformer case have adequate flexibility to accommodate system deflections. In most cases, this can be provided by incorporating a simple dog-leg bend in the piping (Figure 5-24).

5.7.6 Tertiary Bushings and Lightning Arresters

Tertiary bushings on transformers usually provide low-voltage output that is used for station power.

5.7.6.1 Earthquake Performance of Tertiary Bushings. The tertiary bushings and lightning arrester on a transformer have been damaged because the flexibility of the bus support structures imposed loads on the bushings (Figure 5-75) and because the transformer rocked as a result of soil-structure interaction.

5.7.6.2 Recommended Practices for Tertiary Bushings. Conductor connections to tertiary bushings should be provided with adequate flexibility to accommodate movement of the bus support structure and transformer rocking from soil-structure interaction.

5.7.7 Transfer Busses

A special type of bus, called a *transfer bus,* is often used to facilitate the rapid connection of an existing spare transformer at the site or to connect a mobile transformer that is brought from off site to replace a transformer that must be taken out of service. A transfer bus is a two-conductor bus usually installed above a bank of single-phase transformers. One conductor is for the high-voltage and one for the low–voltage connections. Connections between the transformer and the transfer bus are usually made with prepared jumpers. Some utilities have disconnect switches to quickly switch transformers.

5.7.7.1 Earthquake Performance of Transfer Busses. The transfer busses are typically supported on top of a bus support structure over the transformers;

FIGURE 5-75. Tertiary Bushings Were Damaged When the Bus Attempted To Restrain a Flexible Bus Support System.

thus, they are subject to the amplified motion of the tall bus support structure. There have been several failures of these busses. Figure 5-76 shows the transfer busses for two transformer banks. The high-voltage busses are undamaged, but post insulators supporting the low-voltage busses failed. On the far bank, the bus can be seen on top of the bus support structure. In general, the transfer bus is not energized, and its failure would not cause disruption. Unfortunately, in some cases, the failed bus falls across the three phases, shorting them together.

5.7.7.2 ***Mitigation and Retrofit of Transfer Busses.*** When transfer busses fail, damaged parts that disrupt service are removed and service is quickly restored. The bus is then restored when it is convenient. The strength of post insulators used to support the bus should be capable of withstanding the amplified response of the tall support structure. Some utilities have removed their transfer busses, feeling that temporary conductors can be installed quickly when needed.

5.7.8 Emergency Response Procedures for Transformers

5.7.8.1 ***Temporary Repairs for Bushings.*** To make a temporary repair of a leaking bushing to get the transformer back in service quickly, the oil

FIGURE 5-76. The Low-Voltage Transfer Bus for Both Banks of Transformers Failed. *Note:* The bus on the far bank can be seen lying on top of the bus support structure.

around the base of the bushing is removed with a solvent, and fast-setting epoxy is applied to a thin wrap of linen around the porcelain-flange interface. If the leak persists, a vacuum line can be attached to the top of the bushing, and the cleaning and epoxy process repeated one or more times in an attempt to stop the leak. Bushings repaired in this way are later removed and repaired when service conditions permit.

There have been several cases where it was important to keep a transformer in service, even though the bushing was leaking. For smaller leaks, the transformers have been kept in service, and the oil level in the bushing closely monitored.

5.7.8.2 Stopping Radiator Leaks. Radiator leaks can frequently be stopped by tightening the bolts on the leaking flange. In some cases, the transformer was kept in service with the leak until service conditions allowed the leak to be repaired.

An innovative method to temporarily stop a leak was installed by line maintenance personnel. The leaking radiator connection was covered by

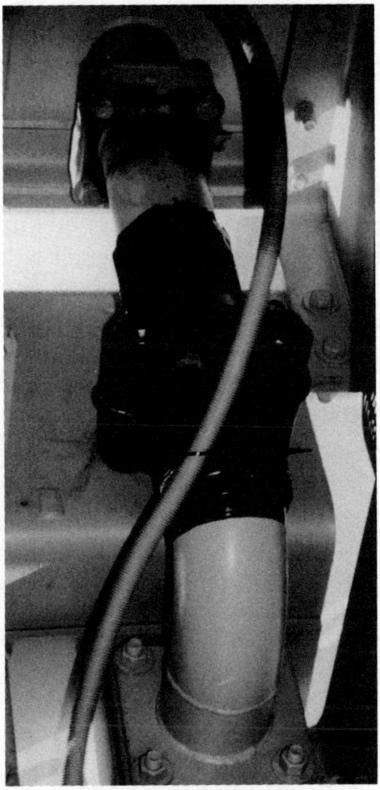

FIGURE 5-77. A Boot Normally Used To Protect Buried Cable Was Used To Seal a Leaking Flange.

an adjustable boot, normally used to protect buried cable connections (Figure 5-77).

At one substation where a manifold radiator was leaking, the flange on the upper pipe connecting the radiator to the transformer was so deformed that tightening the bolts would not stop the leak. To put the transformer back into service quickly, a center punch and hammer were used to deform the flange to stop the leak.

5.7.8.3 Stopping Oil Leaks in Transformer Cases. A hole was torn in the transformer case where it had been welded to an embedment. To keep the transformer in operation until other system repairs could be made, a wooden spike was carved to fit the hole and driven in place to stop the leak.

A small leak at a weld in a transformer case was stopped by using a center punch and hammer to deform metal near the leak to seal it.

5.7.9 Summary of Earthquake Issues Related to Transformers

Because of the large number of issues and the complexity of some of the issues related to the retrofit and installation of transformers, a checklist of issues is given below.

5.7.9.1 Sudden Pressure, Bucholtz, and Protective Relays.

- Earthquakes may cause sudden pressure and protective relays to trip.
- Before resetting sudden pressure relays and putting transformers back into service, the potential for internal damage should be considered.

5.7.9.2 Anchorage.

- The foundation slab must be thick enough for embedments to develop the needed pull-out strength.
- The area of the foundation slab or the use of piles must be such that the bearing capacity of the soil is not exceeded. Where appropriate, the soil under the transformer should be compacted prior to pouring the foundation; poor soils may need to be replaced with engineered fill.
- The foundation slab dimensions should be designed to withstand overturning moments.
- The foundation slab must be strong enough to withstand bending moments for that part of the slab that extends outside the transformer footprint.
- Where conditions for soil-structure interaction exist, various options to reduce potential impacts should be considered. These include providing generous slack to conductor connections; using seismically robust bushings; and using piers, piles, or other means to control foundation movements.
- Anchorage design should consider the strength and stiffness of the entire load path.
- Welding the transformer to embedded steel is the preferred method of anchorage.
- Embedment plates must be stiff enough to prevent welds from tearing the plate, to distribute load to headed studs or other means of securing the plate to the foundation slab, and to avoid unacceptable thermal deformation during concrete curing.
- Voids in concrete trapped under embedments should be avoided.
- When used for anchoring, studs should be headed rather than "L" or "J" bolts.
- Welds to embedments should be designed to avoid stress concentrations that will cause the weld to unzip.
- Welds should be designed to accommodate differences in stiffness of the members along the weld length. Welds should conform to minimum requirements of AISC or AWS.

- Caution should be used in welding to the transformer case to achieve good penetration without overheating.
- When anchoring to an existing foundation slab, grouted, adhesive, or undercut expansion bolts should be used.

5.7.9.3 ***Bushings.***

- The performance of bushings is dependent on manufacturing design details and possibly on maintenance practices.
- Conductor connections between bushings and lightning arresters should be installed so that lightning arrester failure does not adversely affect bushings. See Section 5.9, Lightning Arresters.
- The use of restraint bands around the porcelain-flange joint may prevent porcelain from slipping and gaskets from being displaced.
- Conductor connections to bushings should be provided with adequate slack and flexibility.

5.7.9.4 ***Radiators.***

- Radiators should be braced to reduce moments on pipe connections that penetrate the transformer case.
- For self-supported radiators, connections between the radiator and transformer should have adequate flexibility.

5.7.9.5 ***Conservators.***

- The load path of conservators should be capable of supporting the conservator.
- Piping connections between the conservator and the transformer case should have adequate flexibility to accommodate the dynamic motion of the conservator.

5.7.9.6 ***Tertiary Bushings and Surge Arresters.***

- The flexibility of tertiary bus connections should be compatible with the flexibility of the tertiary bus and its support structure.

5.7.9.7 ***Transfer Busses.***

- Transfer bus porcelain supports should be able to accommodate the loads associated with the deformation of tall bus support structures.

5.8 DISTRIBUTION TRANSFORMERS

Distribution-type transformers can be found in substations, so they are discussed in this section. Distribution transformers are typically installed using one of five methods. Older pole-mounted transformers were installed by hooks over the cross arms. Today, pole-mounted transformers are bolted to the pole. Some transformers are supported on platforms between two poles or, for larger transformers, the platform may also be supported by intermediate poles. Figure 5-78 shows a pole-supported platform where the dotted lines indicate an optional intermediate support. Most other distribution transformers are supported on concrete slabs at grade, in vaults at grade or in vaults below grade. These transformers typically do not have tabs at their base to facilitate anchoring.

5.8.1 Earthquake Performance of Distribution Transformers

The earthquake performance of pole-mounted transformers that are bolted to their poles has been very good. Platform-supported transformers are often unanchored, although some are restrained by wood side rails. In the 1952 Kern County earthquake, more than 800 elevated transformers fell from their supports (Figure 5-79). It is not known what percentage were supported on platforms or were hooked on crossarms. In many cases, the wood side rails did not restrain the transformers. There have been very few

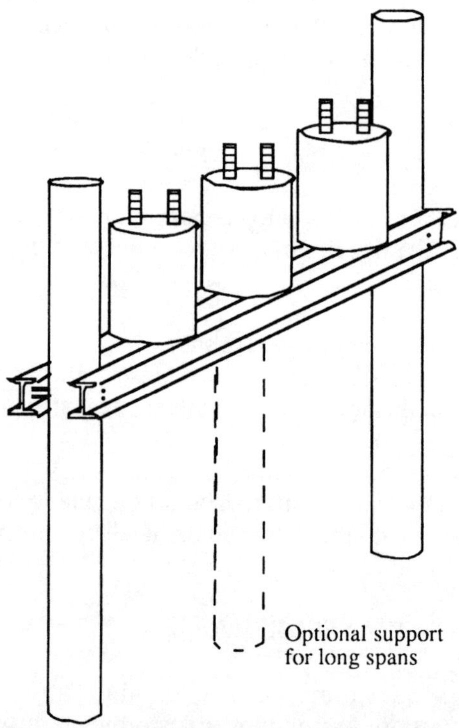

FIGURE 5-78. Pole-Supported Transformer Platform.

FIGURE 5-79. Platform-Supported Transformers Are Frequently Unanchored to Their Supports.

Note: In this picture, one of the platform-mounted transformers was restrained by its conductors, and the others only tilted.

failures due to the failure of wooden poles, even in areas where many poles had severe dry rot. In Japan and the Philippines, there have been many examples of settling of poles when soil liquefied (Figure 5-80). Platform-mounted transformers are used by some utilities in congested urban areas because land for a vault is not available or is very expensive. Should these units fall, they present a hazard.

Most California utilities have not anchored slab-mounted distribution transformers. In the Northridge earthquake, more than 250 slab-mounted transformers moved on the slabs (Figure 5-81). While many transformers moved more than 30 cm (1 ft) and power cables entered the transformer through its foundation slab, there was no record of cable damage as a result of transformer movement. The transformers had to be repositioned, and the utility's policy changed so that new slab-mounted transformers will be anchored.

FIGURE 5-80. The Poles Supporting These Transformers Sank into the Liquefied Soil Several Feet; a Pedestrian Could Touch the Transformers from the Roadway.

5.8.2 Mitigation and Retrofit of Distribution Transformers

The current practice of bolting pole-mounted transformers to their poles has been effective. As remaining crossbar-hook-type transformers are replaced, the vulnerability of this type of transformer will be eliminated. If hooked transformers are used at critical facilities, it is suggested that a few wraps of wire across the open hook or around the crossbar and hook should keep these units in place. Most slab-supported distribution transformers are provided with anchor tabs near their base plates. Since the disruption caused by the lack of anchorage has been minor, retrofitting may not be justified, except at critical facilities. It is recommended that standard practice for new installations provide anchoring.

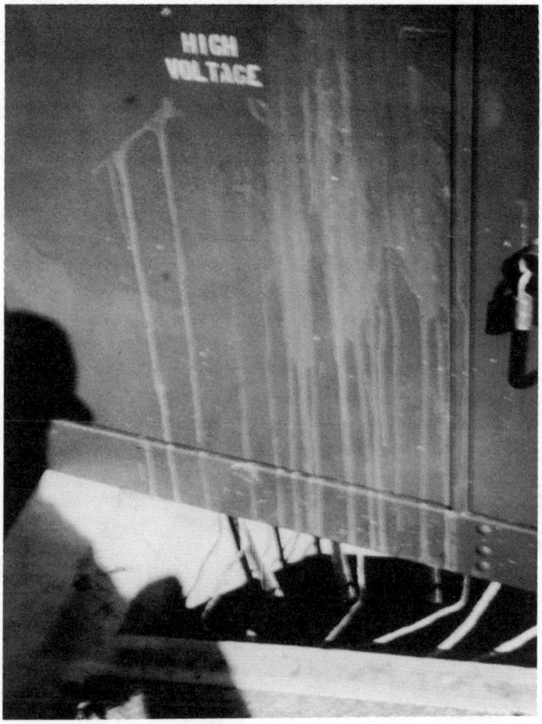

FIGURE 5-81. Slab-Mounted Distribution Transformer Was Unanchored and Slid on Its Slab.

5.8.3 Recommended Practice for Distribution Transformers

Distribution transformers should be anchored to their support structures. In some California utilities, the engineering department recommendations to anchor platform-supported transformers have been ignored by operations, reportedly to facilitate rapid transformer replacement. Figure 5-82 shows a method for restraining platform-mounted transformers that should not impede rapid replacement. For each transformer, a channel is bolted to the platform adjacent to the transformer in the front and back. The figure only shows the front channel, but the inset shows the positioning of both restraints. A wood member that makes contact with the transformer case is secured inside of the channel. Plastic industrial strapping is used to secure the transformer to the restraints (shown separated from the restraints and transformer for illustrative purposes). The restraints are placed at one end of the transformer rather than at opposite ends so that they can accommodate any size transformer.

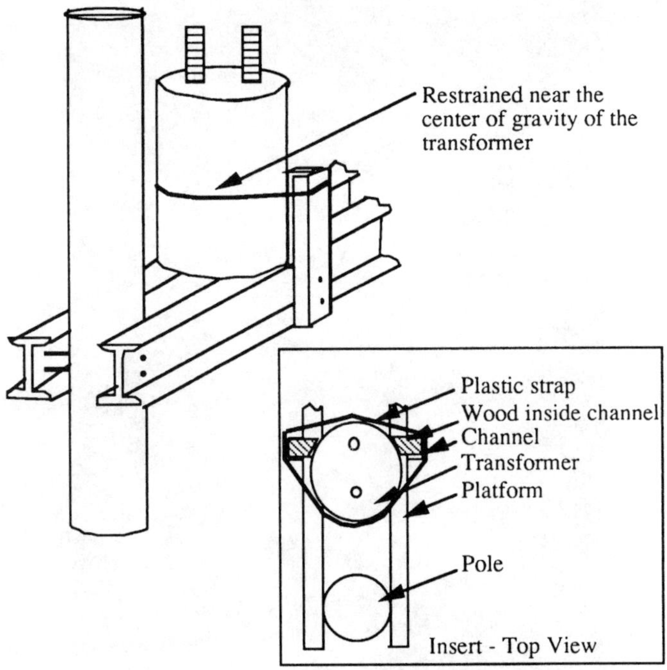

FIGURE 5-82. Method for Restraining Platform-Mounted Transformers to Their Platforms.

5.9 LIGHTNING (SURGE) ARRESTERS

A lightning, or surge, arrester limits the voltage on the line to which it is connected. Under normal conditions, it acts as an open circuit. If the voltage exceeds a predetermined value, the lightning arrester becomes a low-impedance path to ground. Some lightning arresters are fitted with strike counters. These are devices that count the number of times that a lightning arrester is activated. For strike counters to operate, the lightning arrester base is electrically isolated from its support with short standoffs, and the ground lead goes through the strike counter. Lightning arresters and their standoffs are considered in this section.

A lightning arrester consists of a porcelain member with a mounting flange at its base and conductor connection at the top. Lightning arresters made of a composite material rather than porcelain are available, but few high-voltage units have been subjected to earthquake. Some lightning arresters are fabricated by bolting together a stack of shorter units. Lightning arresters are typically connected to the high- and low-voltage leads near transformer bushings. They are supported on top of the transformer, on booms extending from the transformer case, or on a separate post near the transformer. The last is common for equipment operating at or above 500 kV.

5.9.1 Earthquake Performance of Lightning Arresters

Lightning arresters are the most frequently damaged item in a substation. They usually fail at the sand ring just above the mounting flange, or the standoffs fail (Figure 5-83). About equal numbers fail at the sand ring and the standoffs. It has been observed that standoffs are often used even when strike counters have not been installed. Also, it appears that even when strike counters are installed, they are no longer actively used.

The failure of lightning arresters is due to vibration-induced loading or to interaction loads applied through the conductors. Figure 5-84 shows three damaged lightning arresters on the ground next to the transformer. These lightning arresters were connected to the main conductor by drops from the conductor. (The conductor drops had been removed by the time the picture was taken.) One of the reasons for the high failure rate of lightning arresters is probably due to the amplified response of the lightning arrester support structures supported from the transformer case. Figure 5-85 shows a support boom without a lightning arrester. (This lightning arrester failed at the sand ring; the base plate was removed prior to the picture being taken.) This failure was attributed to the lack of slack between the lightning arrester and

FIGURE 5-83. Failed Lightning Arresters

Note: The two lightning arresters in the front failed at the standoff, while the one further back failed at the base of the body of the arrester.

FIGURE 5-84. Three Failed Lightning Arresters.

Note: The drops from the overhead conductor were removed before the picture was taken.

a disconnect switch. The disconnect switch support structure was relatively flexible.

Most 500-kV and many lower-voltage lightning arresters are mounted on their own support posts near the transformer. Other substation equipment, such as potential transformers, current-voltage transformers, and wave traps have similar supports. There have been some failures of this type of detail. Figure 5-86 shows the failure of the weld between the post and the base plate. Close inspection suggests poor weld penetration. Figure 5-87 shows concrete breakout of an anchor bolt. There have been many cases where the anchor bolts have stretched or pulled out of the foundation slabs. This can be difficult to evaluate because the nuts are often spiked to prevent

FIGURE 5-85. This Lightning Arrester Probably Failed Because of the Lack of Slack between It and the Disconnect Switch to Which It Was Connected.

them from turning or the loads have jammed the nut and bolt so that they do not turn freely. For this reason, it is very useful to place a washer under the nut. If the bolt has stretched, it is easy to check if the washer is loose. One factor that contributes to the dynamic response is flexibility of the anchor detail. It is important to distinguish between the strength of an anchorage and its flexibility. A thin base plate with anchor bolts relatively far from the support column and no gusset plates can constitute a flexible overall system. This tends to lower the natural frequency into the high-energy part of the earthquake spectra and to generate relatively large displacements at the top of the lightning arrester. Depending on the conductor configuration, the restraining loads may be transmitted to the transformer bushing.

5.9.2 Mitigation and Retrofit of Lightning Arresters

The result of most lightning arrester failures is that the transformer shuts down, since the dangling lightning arrester typically causes a short circuit with the transformer case or other grounded member. A potentially more disruptive effect is damage to a transformer bushing. This can occur when the failed lightning arrester swings into the bushing. Falling lightning arresters have also damaged conductor posts on bushings.

In some configurations, it may be possible to reconnect the conductors on the lightning arrester so that it is not connected directly to the bushing; it

FIGURE 5-86. This Post Fell Over When the Weld between the Base Plate and Post Failed.

may also be possible to use a smaller conductor to the bushing. These methods are discussed in detail in the Section 5.9.4 on recommended installation practices.

About half of lightning arrester failures can be attributed to vulnerable standoffs. The earthquake performance can be improved by removing the standoff when strike counters are no longer actively used or by replacing standoffs with less vulnerable designs.

5.9.3 Emergency Response Procedures for Lightning Arresters

Within California, the most common practice is to remove a damaged lightning arrester from the circuit so that the transformer can be put back into service. It is not uncommon for months to elapse before units are

FIGURE 5-87. Loads on a Post Anchor Caused the Bolt To Break Out of the Foundation Because of Inadequate Edge Distance.

replaced. In most areas of California, lightning is very rare most of the year. This practice would be ill advised in areas more vulnerable to lightning.

5.9.4 Recommended Installation Practices for Lightning Arresters

Figure 5-88 shows several alternatives, including the preferred methods, for connecting a lightning arrester to a transformer. Some of the principles can be used for other conductor configurations, including the use of a rigid bus. Case A shows one of the common methods of configuring the conductor. The difference between it and Case E is that the upper connection of the drop to the lightning arrester is moved away from the bushing. The intent of this configuration is that, if the lightning arrester fails, it will swing away from the bushing and be less likely to strike and damage the bushing.

Cases B and C are similar to Case F. It is preferable to connect the lightning arrester conductor to the conductor that drops to the bushing, rather than connecting it directly to the bushing binding post. In the preferred method, Case F, if the lightning arrester fails, the weight of the falling lightning arrester is placed on the conductor drop. This connection will be more flexible, so the impact load will be less and the load will be shared by the upper end of the drop and the bushing. If it is connected to the bushing binding post, all of the load will be applied to the bushing; because this connection is relatively stiff, impact loads will be large. Finally, the size of the lightning arrester–bushing conductor can be lighter weight. This will not

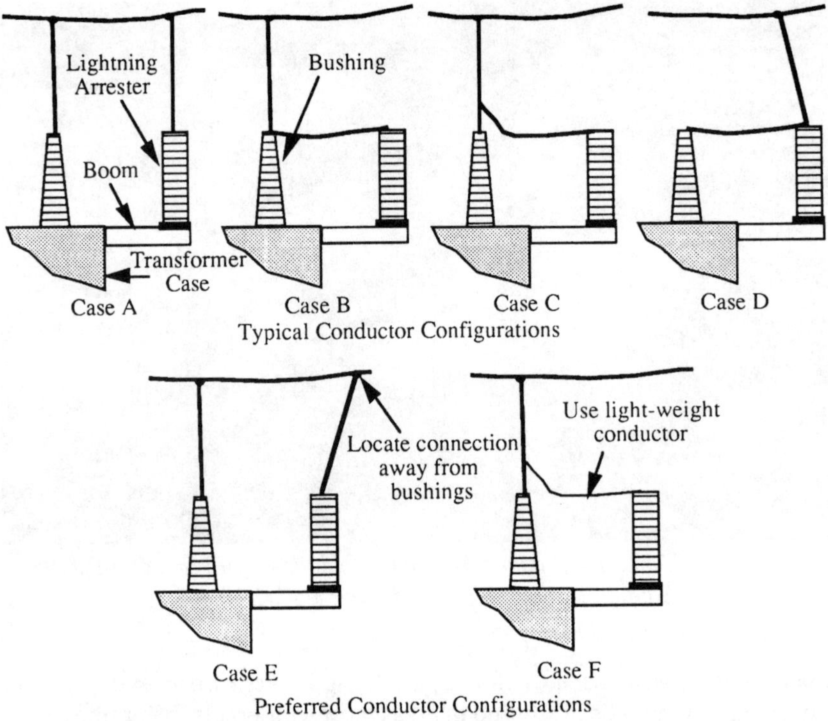

FIGURE 5-88. Variations in Bushing–Lightning Arrester Connections.

affect the operation of the lightning arrester, and the connection may break at a lower force level, reducing the load on the bushing and its conductor.

The least desirable configuration is Case D; the transformer current is carried on bushing–lightning arrester conductor, so its size cannot be reduced and, if the lightning arrester fails, it will tend to swing into the bushing. In some configurations, if the overhead conductors are skewed relative to the line of bushings, there can be a problem of phase-to-phase clearance due to the slack in the lightning arrester drop. This can be addressed by staggering the position of the upper contacts to the overhead conductor. For example, the center phase upper attachment point could be positioned further from the bushing.

The base connection of self-supporting, post-mounted lightning arresters should be stiff. Figure 5-89 shows a very thin base plate, and the bolts are positioned further from the tube wall than is needed. Figure 5-90 is a schematic diagram that illustrates design principles. As the bolts are moved closer to the tube wall, a given moment exerts a larger force on the bolt. If the bolts are moved away from the tube wall, a stiffening gusset should be added between the base plate and tube.

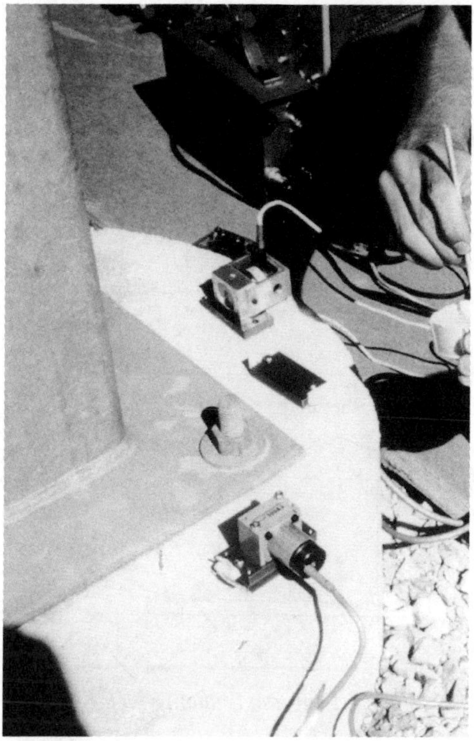

FIGURE 5-89. A Thin Base Plate and Bolts Located away from the Tube Wall Create a Flexible Anchorage System.

About half of the lightning arrester failures are due to the failure of standoffs. They should not be used unless a strike counter is needed. Standoffs have two designs. Those shown in Figure 5-83 consist of porcelain with a stud protruding from each end. When installed and under load, the axial load to resist overturning moments must be carried by the porcelain in tension. Figure 5-91 shows another design. The porcelain members sandwich the anchor tab of the lightning arrester. A bolt passes through holes in the porcelain and anchor plate at the top of the support post. In this configuration, the bolt carries the tensile loads, and the porcelain is under compressive loads. No failures have been observed in this type of standoff.

Generally, seismic design criteria in this document are drawn from IEEE Standard 693. For lightning arresters supported on transformers, the IEEE recommended practice has a safety factor of 2 over that for post-mounted lightning arresters. The flexibility of lightning arrester support booms, when combined with the overall transformer amplification and the possibility of soil-structure interaction, suggests that a more conservative factor of safety should be considered. However, this suggested added conservatism should

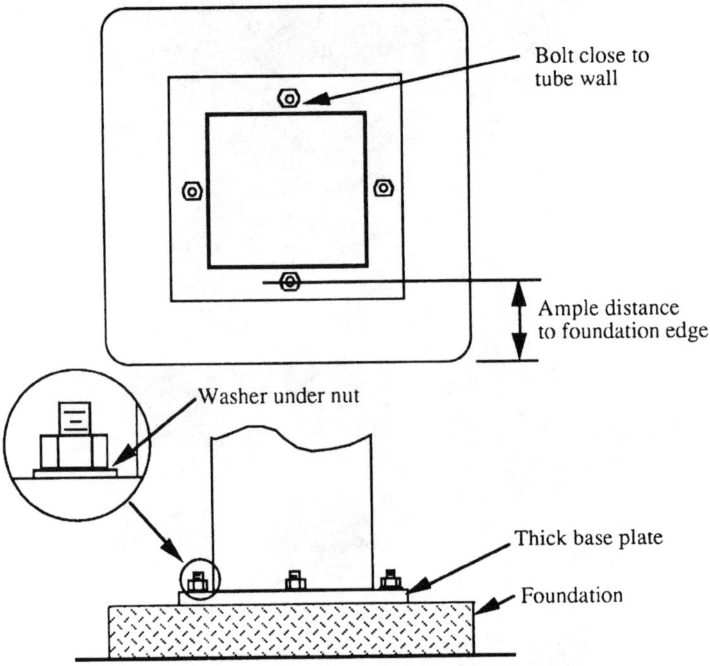

FIGURE 5-90. Schematic Diagram Showing Details for a Stiff Anchorage.

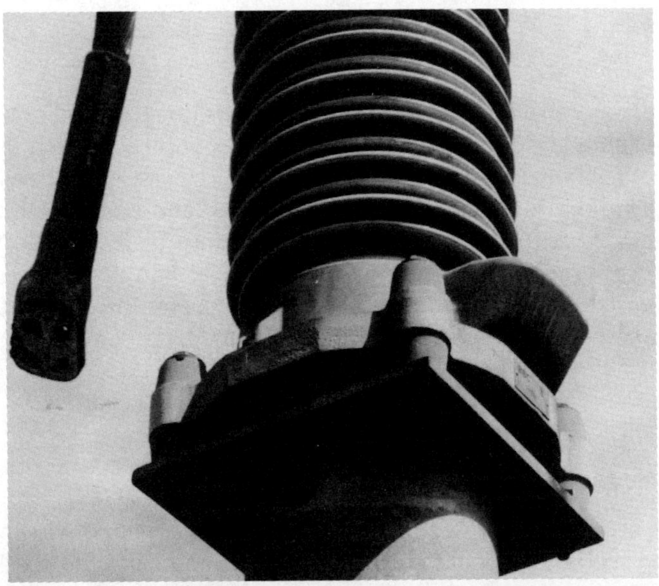

FIGURE 5-91. This Standoff Design Puts Porcelain under Compression and the Bolt Carries Tensile Loads.

be weighed against the relatively small potential for long-term disruption associated with lightning arrester failure if the installation recommendations contained in this document are followed.

5.10 CURRENT TRANSFORMERS

Current transformers are used to measure the current flowing in a high-voltage circuit. The data that they provide is used for system protection, metering, and other control functions. A current transformer consists of a large bushing on top of a box, and it usually has its own support structure. It is frequently located adjacent to circuit breakers. Current transformers are also incorporated in gas-insulated circuit breakers, and some are incorporated in circuit breaker bushings; these types are not considered here but are part of the general performance of circuit breakers.

5.10.1 Earthquake Performance of Current Transformers

Seismic loads on current transformers are due to vibration response of the equipment, including its support structure, and to interaction loads from adjacent equipment through conductor connections. There have been cases where the dynamic response of the current transformer caused damage to the circuit breaker without being damaged itself. As is the case in assessing the earthquake performance of many items of substation equipment, when adjacent items of equipment are damaged, it is difficult to determine which item failed first or whether interaction loads caused the failures. Current transformers are located adjacent to circuit breakers, and there is often interaction between the two items.

Current transformers have developed leaks between the bushings and the supporting box. Some of these leaks appear to have occurred independent of interaction with the adjacent circuit breaker.

Figure 5-92 shows an overview of an undamaged live-tank circuit breaker and current transformer, as well as the lack of slack in the connection between the units. The support structures are relatively stiff, but the tall, slender interrupter-head columns of the circuit breakers are relatively flexible. Figure 5-93 shows that the connections between the circuit breaker and the current transformer failed. A close-up view of the current transformer connection shows that the cast-aluminum cable fitting failed (Figure 5-94). However, Figure 5-95 shows that the interconnection did not break before the porcelain near the flange that joins the upper and lower parts of the current transformer failed. In this case, the flexibility of the circuit breaker structure and the lack of slack in the connection to the current transformer caused the failure of the current transformer. Figure 5-96 shows a current transformer and the rigid connection between it and the

FIGURE 5-92. A Live-Tank Circuit Breaker with Limited Flexibility to the Adjacent Current Transformer.

adjacent circuit breaker. The porcelain strut was provided to resist wind loads. The base of the current transformer is supported on a chair, shown in Figure 5-97. The base of the chair is much larger than that of the current transformer, so the channels provide a flexible support. An evaluation of the support indicates that the natural frequency of a current transformer supported in this way is between 1 and 2 Hz. This places it in the high-energy part of the ground spectrum. The flexibility of the current transformer support structure and the lack of slack in the conductor between the circuit breaker and the current transformer contributed to the failure of the circuit breaker. As a result, two circuit breaker interrupter head support columns failed, as well as many of the porcelain struts. Strong motion

FIGURE 5-93. Failed Connections between Circuit Breakers and Current Transformers.

instruments near the site recorded peak ground accelerations of only 0.05*g*. At the same site, another circuit was installed about a year after the first, using the same equipment, but the chair was smaller so that legs were directly under the current transformer case. This raised the natural frequency of the current transformer, and none of the circuit breakers were damaged in the earthquake.

Figure 5-98 shows damaged current transformers resulting from the failure of live-tank circuit breakers. The conductors seen hanging from the current transformer had been retrofitted to provide flexible connections, but the collapse of the circuit breakers made excessive demands. Note that the conductor hardware failed, but not before damaging the current transformer.

FIGURE 5-94. The Cast-Aluminum Cable Connection Failed.

FIGURE 5-95. Conductor Interaction Loads Probably Caused the Failure of the Porcelain Just below the Flange between the Upper and Lower Halves of the Current Transformer.

FIGURE 5-96. Flexibility in the Current Transformer Support Structure and the Lack of Slack in the Conductor Resulted in Damage to the Circuit Breaker.

FIGURE 5-97. The Base of the Current Transformer That Is Supported on a Chair.

Note: The flexibility in the support channels lowered the natural frequency of the current transformer support structure assembly.

5.10.2 Mitigation and Retrofit of Current Transformers

Providing adequate slack between the current transformers and circuit breakers has been done to prevent interaction and damage. Figure 5-99 shows a flexible link that replaces the rigid connection shown in Figure 5-96.

As can be seen from Figure 5-98, generous slack will not prevent damage if equipment collapses. An alternative is to incorporate a breakaway connection to isolate adjacent pieces of equipment. This is what happened

FIGURE 5-98. Collapse of the Live-Tank Circuit Breakers Damaged the Adjacent Current Transformers.

with the connection shown in Figure 5-94, but in this case the connection was not an engineered breakaway connection and the failure load still caused damage to the equipment. These issues are discussed in Sections 5.6.4 and 5.6.6.

5.10.3 Emergency Response Procedures for Current Transformers

Plastic sheeting and absorbent material is often placed under leaking current transformers to confine the leaking oil and avoid the need to remove contaminated soil. Frequently, the available supply of the normal material used as an oil absorbent is exhausted, and locally available cat litter has been used. Keeping a roll of plastic sheeting at substations can facilitate a fast response and reduce oil contamination.

FIGURE 5-99. A Flexible Connection between the Circuit Breaker and the Current Transformer Has Replaced a Rigid Connection That Contributed to Circuit Breaker Damage.

The failure of current transformers is usually accompanied by circuit breaker failures, so bypassing the current transformer will not solve the problem. When an isolated current transformer has been damaged, it is usually part of a breaker-and-a-half bus configuration, and the current transformer can be isolated without disrupting service. In cases of severe damage to the switchyard, the entire switchyard has been bypassed.

5.10.4 Recommended Installation Practices for Current Transformers

The recommended practice is to provide stiff support structures and generous slack to connections with the current transformer. If breakaway connections become commercially available, they could serve to isolate damage. For new designs, equipment should meet the requirements of IEEE Standard 693, so that individual equipment items will be much more robust. Finally, in California, there has been a tendency to use dead-tank circuit breakers, so the need for independent current transformers is reduced.

5.11 INSTRUMENTATION TRANSFORMERS

Voltage-measuring devices take several forms and are referred to by various names, including current-voltage transformers, potential transformers, capacitive coupling potential transformers, and capacitive coupling potential devices. For simplicity, all of the devices will be referred to as instrumentation transformers (ITs). From an earthquake perspective, their external physical form and earthquake performance are similar and will not be distinguished from each other here. Historically, potential transformers tend to be larger and heavier than the other devices, so they may have lower natural frequencies. The IT usually consists of a porcelain column attached to the top of a metal box. The external physical design of units from different manufacturers varies, and their earthquake performance may also differ.

ITs are used to measure line voltage, and their output can be used for system protection, metering, and control functions. The output from many ITs is used to locate line faults, and their loss may reduce the quality of system protection.

The method of installation is influenced by their function and position in the switchyard. ITs are typically supported in one of four ways as illustrated in Figure 5-100. Case A illustrates a column support structure, where the strength and stiffness of the support column can have wide variations at different facilities. Case B illustrates a suspended system; various methods are used to restrain suspended ITs from below. Restraint configurations would include a fixed restraint to a lower support structure, a spring-loaded restraint to a lower support structure, a cable with or without a spring anchored to a foundation slab, or a weight with a guide just below the IT. Case C illustrates a common configuration found below transmission lines where they enter the substation site. Case D illustrates a support attached to adjacent disconnect switches. There are wide variations in design and performance within each of these general forms.

5.11.1 Earthquake Performance of Instrumentation Transformers

ITs operating below 220 kV have performed well. Units operating at 220 kV and above and conforming to the configuration shown in Case A frequently fail in their porcelain where it joins the box, as shown in Figure 5-101. Most failures are isolated, so it is clear that the IT is vulnerable, rather than pulled down by an adjacent item of equipment. In general, units supported on stiff support structures did not fail. In the Northridge earthquake at four severely shaken substations, units with the common, more flexible support column had failure rates of about 15% at three substations and 25% at another. No ITs supported on stiff support structures failed. Belleville washers had been placed on 12 units at one severely shaken substation. None on stiff support columns failed, while four of six on flexible support

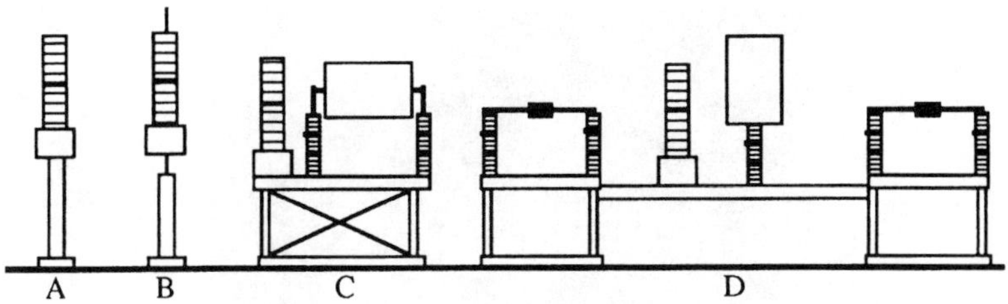

FIGURE 5-100. *Installation Configurations for Instrumentation Transformers.*

FIGURE 5-101. *Instrumentation Transformers with Failed Porcelain.*

Note: Four of six potential transformers supported on more flexible support columns and with Belleville washers failed. The less massive current-voltage transformers supported on stiffer columns and with Belleville washers were undamaged.

columns failed. Thus, the failure rate of the units with Belleville washers on flexible supports was more than twice that of units without Belleville washers on flexible supports.

The performance of units conforming to the configuration shown in Case B is interesting; the survival rate of the ITs is very good, but the equipment adjacent to them appears to have been damaged by interaction loads. Figure 5-102 shows the spring used to restrain the bottom of the IT at the top of the restraint column. (The rope is a temporary restraint.) The buckling and the tearing of the aluminum angle that holds the restraint shackle can be seen. Once the units broke free, the IT moved enough to pull on and tear the insu-

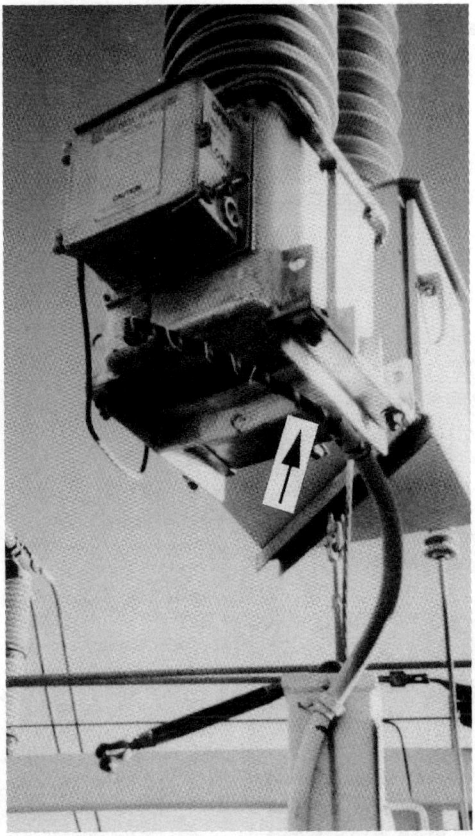

FIGURE 5-102. The Restraint Spring, the Damaged Restraint Attachment Point, and Damaged Signal Cable Illustrate the Damage to Suspended ITs.

lation of the lower-voltage signal cable. The ITs appeared to be electrically undamaged, except for the damaged signal cables. The anchorage on many of the restraint posts was severely damaged, suggesting that they were under high lifting loads (Figure 5-103). The dynamic response of the system is not understood, but it appears that the vertical dynamic response of the structure supporting the IT may have contributed to the damage. The evaluation of the response from two earthquakes indicates that suspended equipment performs better than the same equipment supported on posts. However, it is difficult to prevent the equipment from swinging, so interaction loads can be very destructive to adjacent equipment.

Failures are frequently observed in ITs that conform to the configuration shown in Case C. Figure 5-104 shows a failed IT from which the cables had been removed prior to the picture being taken. The unit had been cleared from the cable. The conductor configuration shown on the undamaged

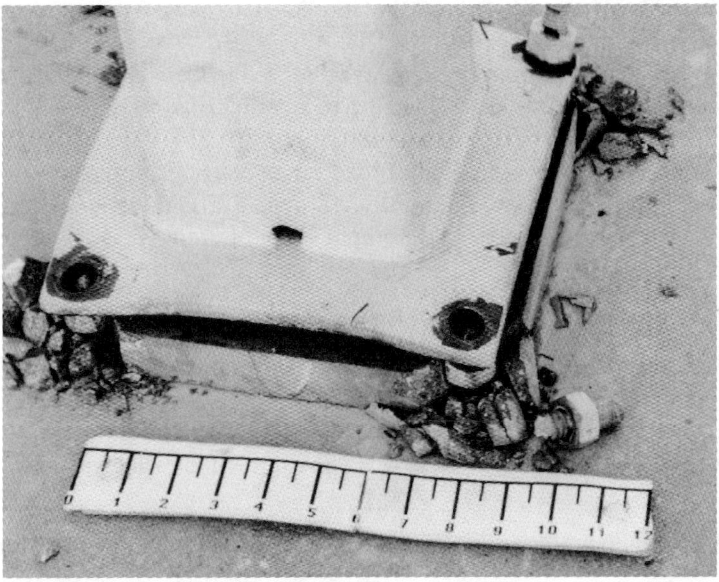

FIGURE 5-103. The Failure of the Anchor Bolts and the Deformation of the Base Plate Indicate the Anchorage Was Subjected to Very Large Loads.

FIGURE 5-104. The IT in the Foreground Has Failed.

Note: The configuration of the conductors seen in the background suggests that the failure was due to inertial loads.

units in the background suggests that inertial loads caused the failure, rather than cable interaction.

The variation of support systems represented in Case D precludes drawing general conclusions about performance, other than to note that failures have been observed for this class of configuration.

The observed failures of ITs in a broad range of installation configurations suggests that the units are inherently vulnerable. The damage pattern suggests that the performance of equipment from different vendors differs.

5.11.2 Mitigation and Retrofit of Instrumentation Transformers

Mitigation methods were attempted by mounting the units on Belleville washers in order to introduce damping into the system. The results were mixed and appeared to be more a function of the stiffness of the support columns and its anchorage than due to the use of the Belleville washers.

5.11.3 Emergency Response Procedures for Instrumentation Transformers

Typically, damaged units are cut from the line, and the system put back into service. Damaged units are replaced as conditions permit.

5.11.4 Recommended Installation Practices for Instrumentation Transformers

The requirements of IEEE Standard 693 should improve the performance of vulnerable ITs. Damage patterns indicate that stiff support structures improve the earthquake performance of ITs. The overall performance of suspended ITs and adjacent equipment indicates that an improved method of restraining these devices is needed. The approach used for suspended wave traps, which is discussed in Section 5.15, is not applicable because of the different requirements for conductor connections.

5.12 CIRCUIT BREAKERS

Circuit breakers are designed to open a circuit when it is energized, even when the current exceeds its normal operating value, for example, when there is a fault on the line. They are designed to operate quickly, from 1 to 5 cycles of line frequency. They are a critical element in the protection system. Various substation bus configurations incorporate different degrees of redundancy, so that one or more circuit breakers, depending on the bus configuration and the position of the circuit breaker, can be taken out of service without affecting the operation of the substation.

Circuit breakers fall into four groups: bulk oil; live-tank, nonpuffer type; live-tank, puffer type; and gas-insulated, dead-tank type. For simplicity, live-tank, nonpuffer type circuit breakers will be referred to as live-tank circuit breakers, and live-tank, puffer type circuit breakers will be referred to as puffer circuit breakers. Some dead-tank circuit breakers use a puffer design; they are grouped with dead-tank circuit breakers and should not be confused with live-tank, puffer type circuit breakers. The earthquake performance, mitigation, and retrofit of each of these types of circuit breakers are described in the following section.

Live-tank and puffer circuit breakers either have a current transformer incorporated into one of the interrupter head columns or have a self-contained unit adjacent to the circuit breaker. As noted in Section 5.10, Current Transformers, interaction between circuit breakers and the freestanding current transformers has caused circuit breaker damage.

5.12.1 Earthquake Performance, Mitigation, and Retrofit of Circuit Breakers

5.12.1.1 Bulk Oil Circuit Breakers. Bulk oil circuit breakers are a type of dead-tank circuit breaker, available at voltages of 345 kV or less. Each phase usually takes the form of a vertical cylindrical tank with bushings on top, although "pocketwatch"-shaped units also exist. Three single-phase units are usually welded to a skid that rests on a concrete slab. The skid can be welded to steel embedments in the slab, bolted to the slab with cast-in-place bolts or expansion bolts, or restrained with bolted friction clips.

In some substations, isolation and bypass disconnect switches are mounted on the bus support structure above each circuit breaker (Figure 5-105).

5.12.1.1.1 Earthquake Performance of Bulk Oil Circuit Breakers. The performance of bulk oil circuit breakers has been very good, with no failures in the United States. Circuit breaker assemblies have moved slightly as a result of slipping friction clips, and anchor bolts have stretched, but movement of bulk oil circuit breakers has not contributed to any damage.

There has been secondary damage to bushings when disconnect switches located above the circuit breaker failed, fell on the circuit breaker, and damaged sheds on bushings (Figure 5-106). In the 1995 Kobe, Japan, earthquake, there were several cases where the bushing porcelain slipped on its support flange, similar to that observed on the transformer shown in Figure 5-60. The differences between U.S. and Japanese practices that would account for the differences in performance could not be determined. There have been many cases where inadequate slack between a circuit breaker and an adjacent disconnect switch has caused a failure, but the disconnect switch rather than the circuit breaker bushing failed (Figure 5-17).

There have been leaks in low-voltage (34.5-kV) circuit breakers where temporary leaks developed between the lid and tank during the earthquake (Figure 5-107).

FIGURE 5-105. Bus Support Structure Supporting Disconnect Switches above Circuit Breakers.

5.12.1.1.2 Mitigation and Retrofit of Bulk Oil Circuit Breakers. To prevent the movement of equipment held by friction clips, the clips are often welded to the skid (Figure 5-108).

***5.12.1.2 Live-Tank Circuit Breakers*.** Live-tank circuit breakers typically consist of one or more metal tanks that contain the interrupter-head mechanism, supported on porcelain columns. Circuit breakers operating between 161 kV and 230 kV typically have two columns. Units operating at 500 kV may have from three to five columns.

The interrupter heads are supported by three different systems. One type consists of porcelain members with flanges at each end, and the column is fabricated by bolting the members together. The second type of construction uses porcelain cylinders that are stacked on top of each other, sepa-

FIGURE 5-106. Secondary Damage to Bushings Caused by Failed Disconnect Switches

Note: Parts of damaged disconnect switches located above the circuit breaker fell and damaged the circuit breaker bushing. The chipped shed did not affect the operation of the circuit breaker.

rated by gaskets and secured by internal tendons that are tightened to hold the column together. In one of these designs, there are internal wood tendons and external diagonal porcelain struts that brace the column for wind loads. A seismically upgraded version eliminated the struts and replaced the wood with fiberglass tendons. A third type of interrupter head uses a space frame made of porcelain and aluminum members. Multiple interrupter head units can be connected with rigid links, flexible braided conductor, or thin stacked aluminum plates. Typically, each phase of the circuit breaker has its own foundation slab and support structure.

5.12.1.2.1 Earthquake Performance of Live-Tank Circuit Breakers. The earthquake performance of live-tank circuit breakers operating at 230 kV and above has generally been very poor. Several failure modes have been observed. Even for a given manufacturer, the form of the circuit breaker has changed significantly as the designs evolved, and each form has its own failure modes. Some of these designs have failed at ground motions as low as 0.05*g*.

Several failure modes have been observed. Some units had anchorage failures and the entire circuit breaker tipped over (Figure 5-109). While other units at this site did not tip over, the circuit breakers were still severely dam-

FIGURE 5-107. Leaks Developed in Low-Voltage Circuit Breakers between the Lid and Case.

aged. The failure of the anchorage on the units that did not tip over could have contributed to the circuit breaker damage. This will be discussed below. The use of fiberglass tendons inside the interrupter-head support columns has better seismic specifications than the use of wooden tendons; however, their earthquake performance was still poor.

Two failure modes are associated with interrupter-head support columns: failure of porcelain members and leaking of gaskets separating column members. The failure of a porcelain support column is shown in Figure 5-110. These columns are often under internal gas pressure. When the column fails,

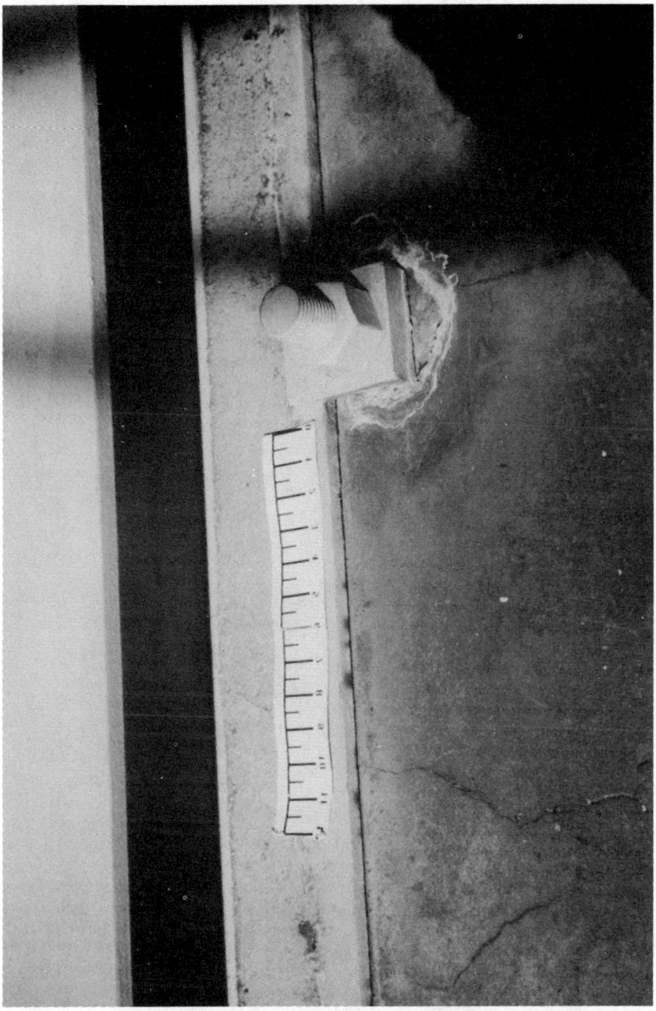

FIGURE 5-108. Bulk Oil Circuit Breakers Are Frequently Held by Friction Clips.

Note: To improve earthquake performance, clips have been welded to the skid as shown here. Because of the large number of clips used to secure the circuit breakers, the slipping or loosening of the clips has not caused any malfunctions.

it does so with explosive force, hurtling porcelain shards up to a 100 ft from the circuit breaker. While this has not caused any injuries to personnel, nearby porcelain has been damaged. The substation shown in Figure 5-110 had four circuit breakers (12 phases) of the type shown in Figure 5-106. This was the only one of the twelve phases that had friction clip anchors that had lifted (Figure 5-111). While it is not known whether the bolt stretched or pulled out, the gap above the clip means that the breaker could have experienced an

FIGURE 5-109. A Live-Tank Circuit Breaker Slid and Tipped Over.

Note: Friction clip anchors proved ineffective to prevent this.

FIGURE 5-110. Failure of a Porcelain Support Column.

Note: One of the three interrupter head columns has failed and fallen to the ground. The fallen column has already been removed. This unit is a seismically upgraded version of the type shown in Figure 5-92, as indicated by the absence of porcelain stays.

FIGURE 5-111. Friction Clips That Lifted.

Note: The friction clips used to secure a circuit breaker allowed the circuit breaker to lift off of the foundation slab. Impact loads when the unit fell back to the slab could have contributed to the porcelain failure.

impact load when the leg dropped back to the foundation slab; this could have contributed to the porcelain failure. The circuit breaker would also have been susceptible to sliding from beneath the friction clip and tipping over.

In this configuration, friction clips have several characteristics that can contribute to damage. First, the clip is held by a single bolt, so that the clip can rotate around the bolt and release the equipment when subjected to a transverse load. Second, if there is a gap between the equipment flange and the bolt, the equipment can slide toward the bolt, allowing the flange on the opposite side of the equipment to slide from under the friction clip. Third, if the equipment slid and is stopped by the bolt, all lateral restraint is provided by that bolt, and the bolt on the opposite side carries no load. Also, if the equipment slides and is stopped abruptly by the bolt, impact loads are introduced. Finally, when subjected to vertical loads that start to lift the equipment, prying action can increase the load on the bolt by a factor of two or more. All of these deficiencies suggest that friction clips should not be used unless they are welded to the flange that is being clamped.

The second failure mode of the support column is that gaskets between porcelain segments are blown or squeezed out, so that the internal gas pressure is lost. The loss of gas pressure may cause an interlock to prevent the circuit breaker from opening. Replacing leaking gaskets can require one or two days of effort, assuming replacement parts are available.

The failure of the interrupter-head gaskets is very common for the circuit breaker shown in Figure 5-112. Two types of failure have been observed as

FIGURE 5-112. One of the More Vulnerable Models of Live-Tank Circuit Breaker.

Note: This site was designed for 0.5*g*. It was constructed before the 1971 San Fernando earthquake, and only circuit breakers "qualified" to 0.2*g* static analysis were available at that time.

the two heads rock back and forth. Note that each column is structurally different, since one column incorporates a current transformer that is contained in the tank below the left column. As a result, their dynamic responses will be different. If they vibrate perpendicular to the plane of the columns, the interaction load causes the gaskets to blow on opposite sides of the interrupter heads (Figure 5-113). If they vibrate in the plane of the columns, the gaskets blow on the top and bottom of the interrupter heads. In looking through the graveyards of damaged equipment after earthquakes, gaskets are blown on units that have collapsed, indicating that the threshold of damage for this type of failure is very low. When there is a loss of pressure in the interrupter head, an interlock prevents the circuit breaker from open-

FIGURE 5-113. An Interrupter-Head Gasket Can Be Seen Protruding along the Side of the Head.

ing. A more flexible connection between the interrupter heads may provide a small improvement in earthquake performance.

Figure 5-114 shows a damaged live-tank circuit breaker. In addition to the damaged support columns, the small diameter, T-shaped porcelain structure has also failed. This carries high-pressure gas to the interrupter head; it is very vulnerable and typically fails before the support columns do. A strong motion instrument in the basement of an adjacent pumping plant recorded a peak acceleration of 0.08*g*. The circuit breakers probably experienced a slightly higher acceleration. The damage pattern in this switchyard was very interesting:[4] the switchyard was constructed on fill across a small ravine, and the units above the deepest part of the fill were undamaged. This switchyard also contained a base isolated circuit breaker (Figure 5-115). While the base-isolated unit was undamaged, nearby units were also undamaged, so this earthquake did not demonstrate the effectiveness of this method of protection. Live-tank circuit breakers of many different designs have been damaged (Figure 5-116).

The live-tank circuit breaker shown in Figure 5-117 is an unusual design, in that a space frame constructed of porcelain and aluminum members supports the interrupter head. Note that the porcelain members are relatively thin. In this configuration, porcelain members are primarily subjected to axial loads rather than bending loads, a much more efficient loading for porcelain. An interrupter-head support on this circuit breaker failed, but it was an aluminum casting rather than porcelain. Figures 5-118 and 5-119 show the interrupter head and the failed aluminum casting. The brittle failure of the casting may have started at a hole in the casting that acted as a stress raiser.

FIGURE 5-114. Damaged Live-Tank Circuit Breaker with Damaged Support Columns and High-Pressure Gas-Piping System.

Figure 5-120 shows a circuit breaker that had been disassembled, moved to the site, and reassembled to rebuild the site. Review of Figure 5-120 shows that the interconnection of the interrupter heads is rigid, and, when the system vibrates, significant torsional loads could be applied to the interrupter head and its support column. Very often, when this type of equipment is analyzed, only one support column is evaluated because all of the columns are the same. When this is done, interaction loads between columns are neglected.

Figure 5-121 shows the damaged circuit breaker before it was replaced by the unit shown in Figure 5-120. While the ground motions at the site were large, this circuit breaker had been qualified by analysis based on response spectra anchored at 0.5*g*. The circuit breaker support structure used bolted connections with slotted holes. An evaluation after the earthquake showed

FIGURE 5-115. Base-Isolated Circuit Breaker That Was Undamaged at the Same Location as the Circuit Breaker in Figure 5-114.

Note: The damage pattern at the site does not demonstrate the effectiveness of this type of protection.

FIGURE 5-116. Live-Tank Circuit Breaker with Interrupter-Head Support Column Damage.

FIGURE 5-117. Live-Tank Circuit Breaker That Uses a Space Frame To Support the Interrupter Head.

that the zinc protective coating around the holes had been burnished by the back and forth action of the bolt in the slot. When a bolt is suddenly stopped at the end of a slotted connection, impact loads are introduced to the structure and the equipment it supports. This may have contributed to the equipment damage.

Many of the various designs of live-tank nonpuffer circuit breakers have been damaged. While some of these units may fail at peak accelerations below 0.1*g*, others have survived 0.3*g*. Details of anchorage, support structures, slack in conductor connections, and the frequency content of the excitation influence their performance.

5.12.1.2.2 Mitigation and Retrofit of Live-Tank Circuit Breakers. Friction clip anchors not only allow circuit breakers to fall over, but may also contribute to porcelain failure by allowing impact loads to be generated. Welding the friction clip to the flange that it is clamping can eliminate this and other problems.

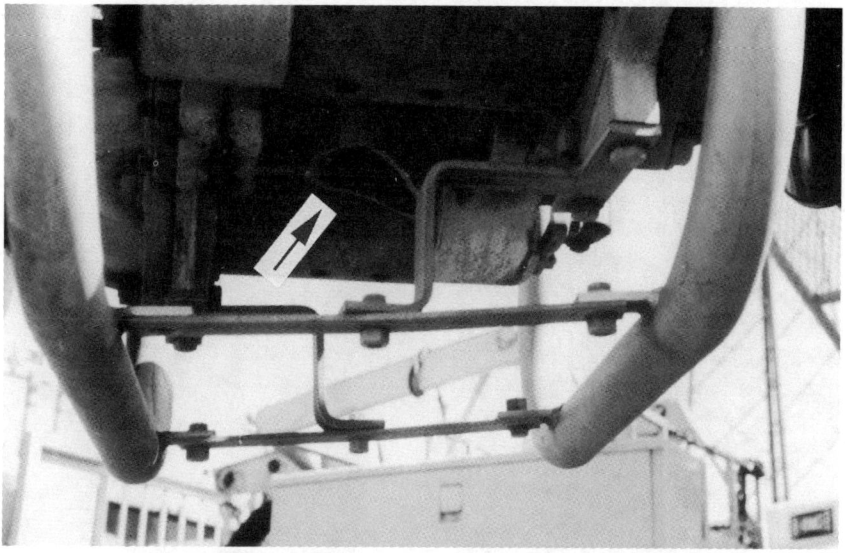

FIGURE 5-118. Interrupter-Head Support Failed.

FIGURE 5-119. Aluminum Casting Supporting the Interrupter Head Had a Brittle Fracture Failure.

FIGURE 5-120. This Circuit Breaker Was Moved from Another Site To Hasten the Rebuilding after an Earthquake.

A suggestion has been made to place a retaining strap around support columns at porcelain joints in an attempt to retain gaskets. While this would be a low-cost remedy, its effectiveness is unknown.

Providing more flexible connections between interrupter heads may improve their performance.

Support structures that utilize slotted connections should have these connections welded after the units are assembled. Figure 5-122 shows a detail of the support structure of the circuit breaker shown in Figure 5-120. The circuit breaker was moved to the site after the earthquake was installed on the existing support structure, and slotted connections were welded to prevent the working of the connections.

FIGURE 5-121. *The Damaged Circuit Breaker before It Was Replaced by the Unit Shown in Figure 5-120.*

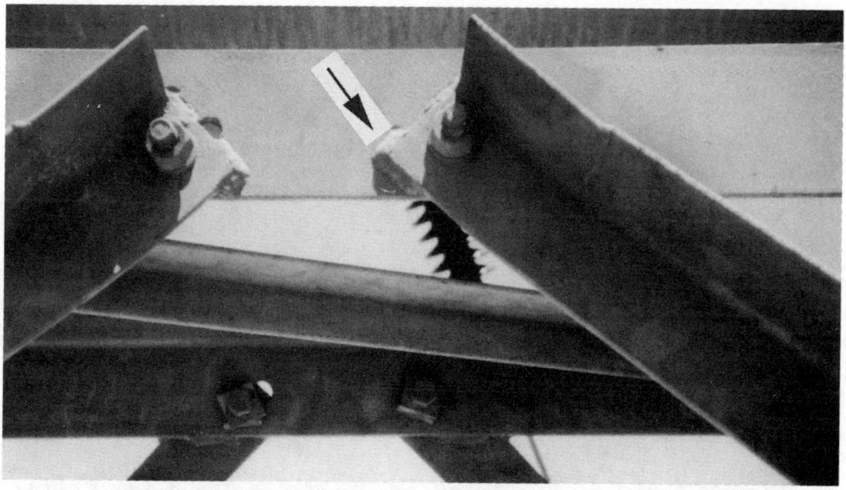

FIGURE 5-122. *A Detail of a Support Structure That Shows a Slotted Connection That Was Welded after the Structure Was Assembled.*

Within California, many of these vulnerable units have been damaged in earthquakes and replaced with more seismically robust units.

5.12.1.3 Puffer Circuit Breakers. Figure 5-123 shows puffer circuit breakers in a 500-kV DC switchyard. In these units, members in the porcelain column that supports the operating mechanism have bolted flanges rather than internal tendons to hold them together. The interrupter mechanism is located at the top of the porcelain column, but the massive tank used in the live-tank design has been eliminated. Some manufacturers have a T configuration, and others have a Y configuration.

5.12.1.3.1 Earthquake Performance of Puffer Circuit Breakers. The introduction of these units into seismically active areas has been relatively recent, so their seismic exposure has been limited; however, the performance of some of these units in strong shaking has been good. There have been some failures. Note that, in the units shown in Figure 5-123, the connection between the circuit breaker and current transformer failed because of inadequate slack; however, only the connection hardware failed, rather than the circuit breaker or the current transformer.

5.12.1.3.2 Mitigation and Retrofit of Puffer Circuit Breakers. The electrical connections to equipment should have adequate flexibility. While each interrupter-head column is structurally the same, minor differences in compo-

FIGURE 5-123. Puffer Circuit Breakers Used in a 500-kV DC Switchyard.

nents or in assembly can cause slight differences in their natural frequency, so connections between interrupter heads should have adequate flexibility.

5.12.1.4 Gas-Insulated, Dead-Tank Circuit Breakers. Most gas-insulated, dead-tank circuit breakers take the form of a horizontal tank with a vertical or near vertical bushing at each end, Figure 5-24. The tank contains the interrupting mechanism and, typically, a current transformer. The tank is at ground potential. These units typically have grouted bushings rather than the post-tensioned variety. The tank is supported on a steel support structure near the ground. Some units are configured with all three phases in a single enclosure, Figure 5-4.

5.12.1.4.1 Earthquake Performance of Gas-Insulated, Dead-Tank Circuit Breakers. These circuit breakers are very rugged, and none have failed except for the one in Kobe noted earlier (Figure 5-4). Subsidence at that site and vertical drops to the bushings caused the bushings to fail.

5.12.1.4.2 Mitigation and Retrofit of Gas-Insulated, Dead-Tank Circuit Breakers. Single-phase units have each phase on its own support structure and may have independent foundation slabs. The units are connected by various control and gas lines. While no problem has been observed yet, some of these connections appear to be vulnerable to relative movement between the phases. In some cases, small-diameter gas lines are routed between phases through trenches in the ground. Cover plates on trenches are often unanchored and in an earthquake the small-diameter tubing could be damaged.

5.12.2 Emergency Response Procedures for Circuit Breakers

Emergency response procedures for circuit breakers include bypassing the circuit breaker and using the next circuit breaker in the circuit for protection. However, there have been cases were the nearest circuit breaker was two substations away.

To replace damaged circuit breakers, units from other sites have been disassembled and reassembled at the damaged site.

5.12.3 Recommended Installation Practices for Circuit Breakers

Circuit breakers should be securely anchored. Adequate flexibility must be provided to all connections to circuit breakers. Connections between interrupter heads on live-tank and puffer circuit beakers should have adequate flexibility. Adequate flexibility should be provided between mechanical and tubing connections of the phases of a circuit breaker. The flexibility provided should account for support and equipment flexibility, and, if phases are installed on separate foundation slabs, movement of the slabs. Small-diameter control tubing should be protected against secondary damage.

5.13 DISCONNECT SWITCHES

Disconnect switches are used to reconfigure a network. They are typically operated when a circuit is not carrying current. In some disconnect switches, nozzles direct compressed air at the break point to extinguish the arc, or resistors are added so that the units can be opened even if the circuit is energized. Disconnect switches are typically found on each side of a circuit breaker, so that the circuit breaker can be taken out of the circuit for maintenance. In some bus configurations, there are also bypass switches around circuit breakers. Disconnect switches are also used to divide sectionalized busses. Switches can be manually or motor operated. Some switches are also configured so that they can ground the circuit on one side of the switch.

Disconnect switches can take many different forms. The most common type found in California is the center-break type, though the vertical-break type also is found. Others seldom seen in California include pantographic, "V", and double-break types. Some switches are supported on simple, unbraced columns so that they are relatively flexible. Others are on elevated structures so that they are more nearly aligned with connections with other equipment, or they are located on top of bus support structures over circuit breakers. Structural amplification of the support structure may increase vibration levels and seismic exposure of the switches. A grounding type switch has an extra link that can ground one side of an open switch.

5.13.1 Earthquake Performance of Disconnect Switches

Several problems have been observed with disconnect switches. The most common failures have been to post insulators. Because disconnect switches are typically located adjacent to circuit breakers, the poor seismic performance of live-tank circuit breakers has often pulled down the adjacent disconnect switch. Frequently, the flexibility of connections between the disconnect switch and adjacent equipment is inadequate, and failures have been observed which are probably due to interaction loads (Figure 5-16). After the earthquake, it is not always clear what caused the damage. If there are two failures, it is difficult to determine which was the primary damage. There have been switch failures where they were used to interconnect sections of long busses. If the sections of the bus support structure are not structurally connected, the relative movement across the switch can cause its failure.

It has been speculated that an open vertical-break switch may be more vulnerable because of the large moment exerted by the open blade of the switch. Since most switches are normally closed, there are no data to support this contention. Center-break switches are very vulnerable.

In Section 5.11, Instrumentation Transformers, it was noted that disconnect switches were damaged due to the interaction with suspended current-voltage transformers. There have been similar problems with suspended wave traps.

Several failures of disconnect switches not related to the fracture of porcelain have been observed. Many of the center-break type switches at a substation had the cast-aluminum plates that support the post insulator deform, so when the switches were operated to the closed position, the contact surfaces would remain open (Figure 5-124). This failure was due to the deformation of the cast-aluminum plate that supported the post insulators (Figure 5-125). Burning of the contacts on closed center-break disconnect switches has also been observed, probably resulting from the movement of the contacts during the earthquake. This, of course, would not occur if the switches were open.

FIGURE 5-124. Center Break Disconnect Switch Contacts Remained Open When the Switch Was Operated to the Closed Position Because the Plate Supporting the Post Insulators Deformed.

FIGURE 5-125. The Deformation of the Cast-Aluminum Base Plate Prevented the Disconnect Switches from Closing Properly.

Disconnect switches are designed with a horizontal structural member that supports the post insulators. In one design, channels form this member. Deformation of the channels caused by inadequate stiffness or inadequate structural ties between the channels has caused the arms of the switch to become misaligned. There have also been cases where the shaft supporting post insulator columns deformed, causing the contacts to be misaligned. Both of these failures were in metal members, rather than in the porcelain.

On a vertical-break type switch, the center post rotates to raise the blade. The weld that connects the shaft to the base plate failed, so the shaft turned without raising the blade (Figure 5-126). Some vertical-break disconnect switches are fitted with resistors to reduce arcing when they are opened. The cast-aluminum hardware that supported these resistors failed at relatively low vibration levels.

An unusual failure was associated with the low cycle fatigue failure of the support shaft of a center-break type switch. A weld holding a water shield created a stress concentration that caused the failure (Figure 5-127). Finally, there have been two types of failures associated with the operating linkage on several types of disconnect switches. In one, a coupling mechanism became deformed, and, in the other, split clamps on a circular shaft used to adjust the switch alignment slipped. These failures caused the switch to malfunction.

In Japan, low-voltage pantographic switches failed. This was attributed to the fact that the two sides of the switch were supported on different struc-

FIGURE 5-126. Vertical-Break Type Switch That Failed Because of the Weld.

Note: The failure of the weld, attributed to poor quality control during manufacture, prevented torque from being transferred from the control to the post insulator. This prevented the disconnect switch from operating.

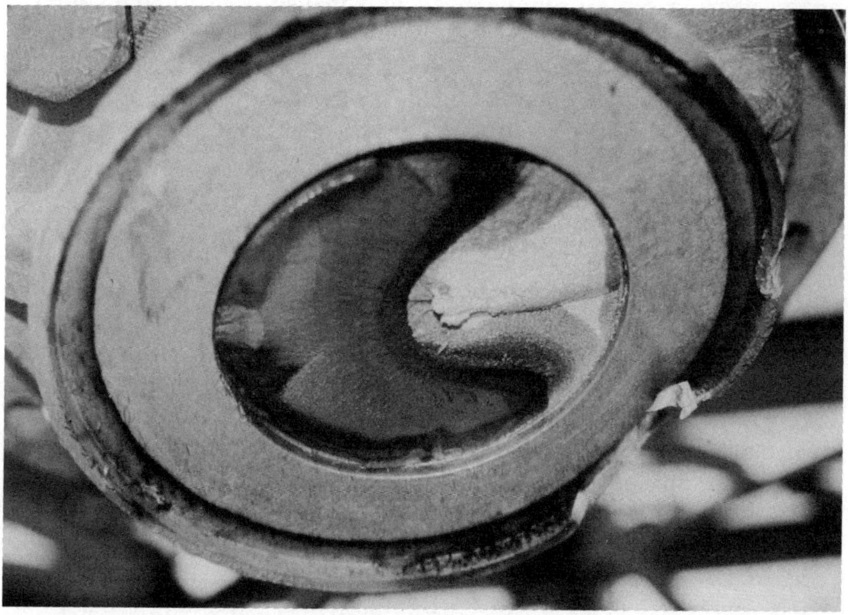

FIGURE 5-127. A Problem in Manufacturing Caused a Low-Cycle Fatigue Failure.

tures and relative displacement between the structures damaged the switches. This would be a vulnerability of most pantographic disconnect switches.

There have been several disconnect switch failures due to poor quality assurance. The most common has been the use of short bolts to assemble equipment, so that only a few threads of the nut were engaged.

Few if any V-type, pantographic, or double-break disconnect switches have had significant earthquake exposure, so their performance is not known.

5.13.2 Mitigation and Retrofit of Disconnect Switches

Providing adequate flexibility to electrical connections of disconnect switches would improve their performance.

In Section 5.15, Wave Traps, mitigation methods to reduce damage from interaction with suspended wave traps are discussed.

Sectionalized bus support structures should have structural members connecting the sections, so that disconnect switches are not subjected to interaction loads.

5.13.3 Emergency Response Procedures for Disconnect Switches

Damaged disconnect switches are frequently bypassed until spare parts are available or service conditions allow damaged units to be repaired or replaced.

To restore service to lines that were disrupted due to the deformation of the switches shown in Figure 5-128, the switches were pulled into a closed position and cotton rope was used to keep the posts together and the switch closed.

5.13.4 Recommended Installation Practices for Disconnect Switches

The importance of adequate flexibility of disconnect switch connections cannot be overemphasized. Many of the failures can be attributed to deficient flexibility.

There were many observed failures not associated with porcelain. The failure of metal parts, the bending of linkages, and the slipping of adjustments all suggest the importance of seismic qualification at levels of expected performance. IEEE Standard 693 establishes a performance level and then recommends that testing be done at half of this level. To assure good performance at the performance level, tests are to be conducted with measurements to assure that the factor of safety will give the desired performance level. This is an acceptable approach when the failure modes are understood and measurements can be made to evaluate stresses, so that fail-

FIGURE 5-128. Cotton Rope Was Used To Keep Deformed Disconnect Switches (Shown in Figure 5-124) Closed So That Service Could Be Quickly Restored.

ures can be predicted and performance assured. In the case of disconnect switches, many of the failures have been unpredictable and are related to nonlinear phenomena, so appropriate measurement cannot be made. For these reasons, it is suggested that disconnect switches be tested at the desired performance level.

5.14 CIRCUIT SWITCHERS

Circuit switchers are used to reconfigure the system. They act like disconnect switches, except that they are designed to operate when the circuit is energized and operating under normal conditions. That is, they are not designed to act as circuit breakers when, for example, there is a short circuit. They are often used to connect capacitor banks to the grid.

A circuit switcher looks similar to a vertical-break disconnect switch, but there is usually a horizontal porcelain member and four vertical members. Some units may have a grounding bar, and this may require an additional porcelain column.

5.14.1 Earthquake Performance of Circuit Switchers

The earthquake performance of circuit switchers has been mixed. At two substations, each with two groups of circuit switchers, one group was severely damaged (Figure 5-126), while the other group had little or no damage. The porcelain columns are the component that usually fail. The units pictured in Figure 5-129 were located about 180 m (600 ft) from the undamaged units. This demonstrates that there can be significant variations in ground motion within a site. There did not appear to be any unusual site conditions that would account for the large difference in motion. This variation in performance was not due to just one circuit switcher, where large variations in porcelain strength could account for the damage. Each location had five switchers, each with three separate phases.

5.14.2 Mitigation and Retrofit of Circuit Switchers

Providing flexible connections may improve the response of circuit switchers.

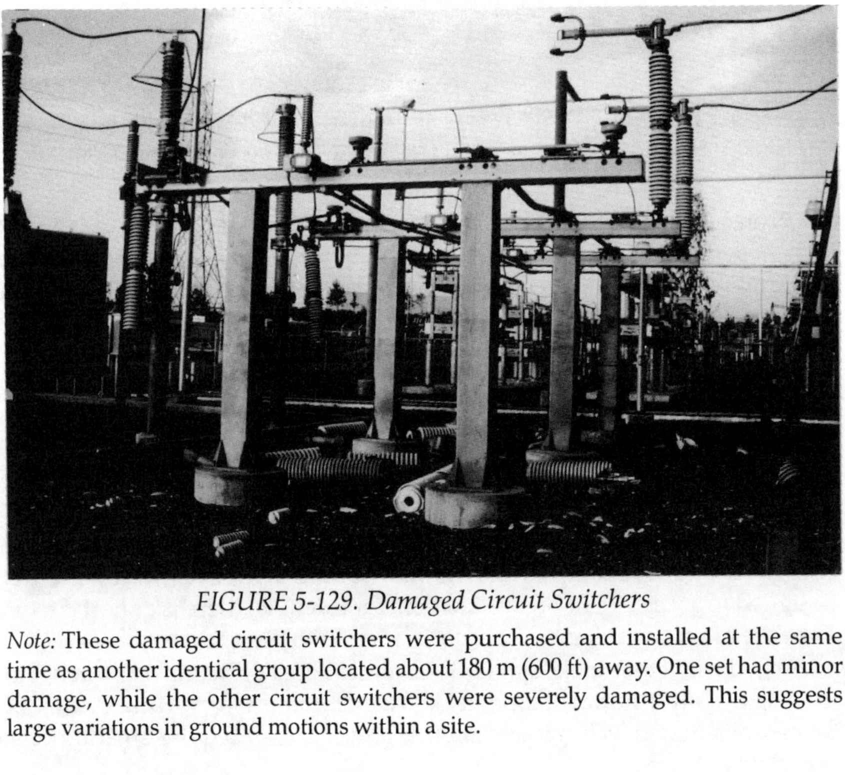

FIGURE 5-129. Damaged Circuit Switchers

Note: These damaged circuit switchers were purchased and installed at the same time as another identical group located about 180 m (600 ft) away. One set had minor damage, while the other circuit switchers were severely damaged. This suggests large variations in ground motions within a site.

5.14.3 Emergency Response Procedure for Circuit Switchers

Figure 5-130 shows temporary conductors used to bypass the circuit switcher. The conductor was supported on remnants of the original columns and on post insulators that replaced some damaged columns.

5.14.4 Recommended Installation Practices for Circuit Switchers

Provide adequate flexibility to electrical connections.

5.15 WAVE TRAPS

A wave trap, or line trap, is an inductor. They are an important component of radio carrier-frequency base communication systems that utilize transmission lines as the medium to transmit information and control signals related to system protection, as well as data related to equipment and system performance. In this system, a high-frequency carrier signal is injected into the power line. To prevent the carrier signal from entering the substation, the wave trap is inserted into the line where it enters the substation. A capacitor-voltage transformer is placed on the line side of the wave trap to provide a low-impedance path for the carrier signal. The wave trap attenuates or

FIGURE 5-130. A Damaged Circuit Switcher Temporarily Bypassed To Restore Service.

blocks the high-frequency signal but has little effect on the 60-Hz power. For lower voltages, there may be a wave trap on only one of the phases. For higher voltages (345 kV and above), it is common to have a wave trap on more than one phase. This also is done on more important circuits for redundancy and to allow the use of more sophisticated protection methods.

It has been standard practice to provide two routes for system protection communication, greatly increasing reliability. This has traditionally been achieved with a line carrier system on transmission lines and on a microwave communication system. With the introduction of optical fiber systems for communications, the use of the carrier system is being eliminated.

Wave traps are usually mounted in one of four ways: supported on a single post insulator with its axis vertical, supported by multiple post insulators (one of which may be a coupling capacitor) with its axis vertical, supported by a post insulator at each end with its axis horizontal, or suspended from flexible conductor with its axis vertical. Suspended wave traps are restrained from below to limit lateral movement in wind and earthquakes. Several restraint methods are used.

5.15.1 Earthquake Performance of Wave Traps

Several failure modes of wave traps have been observed. The most common failure has been the failure of the post insulator support (Figure 5-131).

Several failures have been associated with inadequate structural design or quality assurance (Figures 5-132, 5-133, and 5-134). It is common practice in many utilities to have major structures, such as bus support structures and control houses, designed by civil or structural engineers. However,

FIGURE 5-131. Post Insulator Supporting the Wave Trap Failed.

FIGURE 5-132. A Weld of a Lightweight Aluminum Channel to a Plate Bolted to the Cap of a Post Insulator Failed.

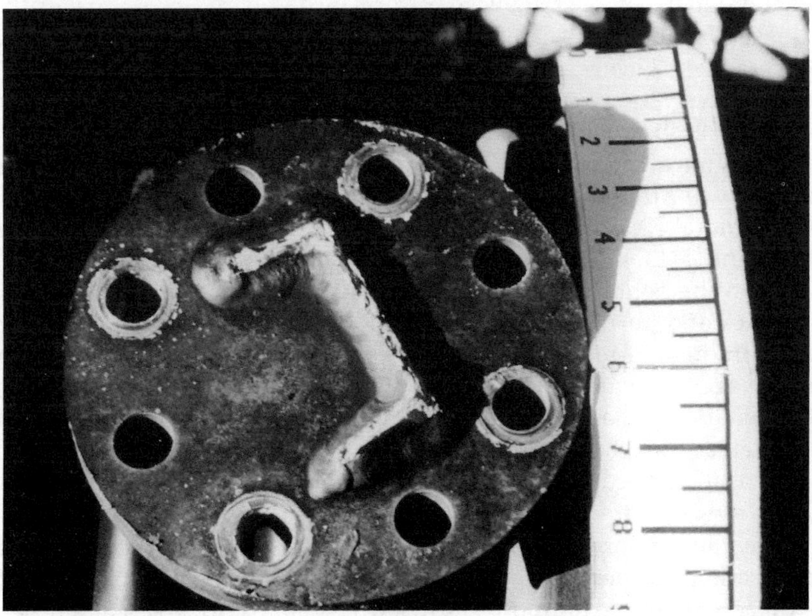

FIGURE 5-133. A Weld to a Plate Bolted to the Top of a Post Insulator Failed.

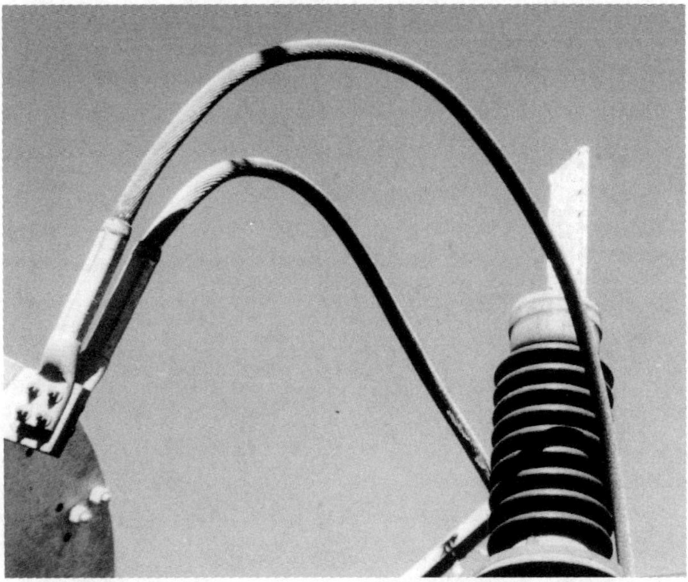

FIGURE 5-134. A Weld on a support leg failed.

many of the minor details, such as support hardware and equipment supports, come from electrical engineering groups. Unfortunately, this has often resulted in poor earthquake performance.

Several methods are used to structurally support suspended wave traps. Some designs have structural members, fabricated from fiberglass, extend from the central core to support the coil that surrounds it. There have been failures of such structural systems within the wave trap (Figure 5-135).

There have been several problems with suspended wave traps. Suspended wave traps with post insulators below the wave trap that provide a stiff lateral restraint have often failed in their lower support (Figures 5-136 and 5-137). These systems had short lengths of cable or chain between the bottom of the wave trap and the top of the lower support. The back and forth motions of an earthquake tended to generate impact loads. In these instances, the restraint failed, and the wave trap itself was undamaged. Frequently, the motion of the wave trap would damage equipment adjacent to it, even though the wave trap would be undamaged (Figure 5-138).

Installation practices can differ in various parts of the country. Figure 5-139 shows a wave trap in which one support leg is provided by a capacitive-voltage transformer. This configuration has not been observed in California, so its earthquake performance is not known, but based on the performance of capacitive-voltage transformers, this would appear to be seismically very vulnerable.

FIGURE 5-135. The Fiberglass Structural System Connecting the Central Core and the Coil Failed.

5.15.2 Mitigation and Retrofit of Wave Traps

Adequate flexibility of conductor connections should be provided. For critical circuits, modifying suspended wave traps as suggested should be considered.

5.15.3 Emergency Response Procedures for Wave Traps

Damaged wave traps are typically bypassed.

5.15.4 Recommended Installation Practices for Wave Traps

Providing adequate flexibility to connections should improve their response. The failures of wave trap internal support structures and also the external support structures will be addressed by the adoption of IEEE Standard 693.

FIGURE 5-136. The Post Insulators Forming the Lower Restraint of a Suspended Wave Trap Failed.

Note: Impact loading probably contributed to the failure.

All equipment supports should be designed by engineers with seismic design experience (see Sections 4.5.5).

The problem with the large number of failures associated with suspended systems remains. If suspended wave traps are installed at the same level as the equipment to which they are connected, it will be very difficult to provide adequate restraint to prevent interaction damage. Three approaches are illustrated in Figure 5-140. In Case A, the stiff short lower restraint tends to restrict movement of the wave trap, but lateral motion would generate impact loads that frequently fail the lower post insulator. In Case B, any relaxation of the tension, such as stretching of the support conductors or deflection of the upper support, will allow relatively large lateral movement of the wave trap. These approaches are not recommended. Case C illustrates the recommended approach. The wave trap is suspended just

FIGURE 5-137. The Aluminum Hardware Connecting the Lower Restraint to the Post Insulators Failed.

below its support structure. It will tend to swing about its upper support but impacts will not be imparted to the upper support. The interconnections to adjacent disconnect switches have little mass, so interaction loads will be small. As a result of the geometry, the movement of the interconnection point due to the swinging of the wave trap will be small.

5.16 CURRENT-LIMITING REACTORS, FILTERS, SHUNT REACTORS, VOLTAGE SUPPORT, AND POWER FACTOR CORRECTION DEVICES

Air core inductors, synchronous condensers, autotransformers, capacitor banks, and static var compensation are used to support voltage or adjust power factor.

FIGURE 5-138. The Lateral Motion of the Wave Trap Damaged Adjacent Equipment through Interaction Loads.

Air core reactors used for these applications tend to be larger than wave traps and are often used in conjunction with other special facilities. They are used on long lines to counteract capacitance effects of the lines, are used to limit power surges at capacitor banks, and are employed in filters at DC converter stations. They are typically mounted with the coil axis in a vertical orientation. The coil is typically supported by insulators. Some units are mounted on insulators at ground level on a concrete slab surrounded by a fence to protect personnel, while others are supported on columns, as in Figure 5-141. The service loads on these devices can be very large. The seismic performance of air core reactors has been good, probably the result of their large service loads plus good seismic design.

Voltage support and power factor correction devices fall into four categories: synchronous condensers, autotransformers, capacitor banks, and static

FIGURE 5-139. Wave Trap Partially Supported by a Current-Voltage Transformer.

var compensation facilities. This equipment is used to stabilize system voltages due to sudden changes in loads or to correct the power factor. At DC converter stations, capacitors are an integral part of filters (dry-type air core reactors being the other main element).

Synchronous condensers are similar to large motors and are becoming relatively rare. In earthquakes, a synchronous condenser's bearings have been damaged, and the pressurized nitrogen system used with these systems has developed leaks (Figure 5-142).

Autotransformers are a special type of transformer, and the discussion on transformers applies to them. They are usually smaller than large power transformers and have small radiators. They differ from regular transformers in that the input and output bushings are the same size.

Capacitor banks usually take one of three forms. The first consists of aluminum racks ranging in height from about 1 m to 4 m (4 ft to 12 ft) and sup-

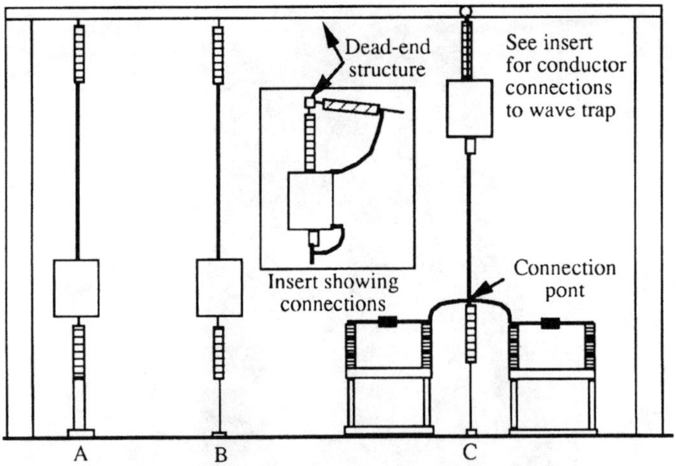

FIGURE 5-140. Methods of Installing Suspended Wave Traps.

FIGURE 5-141. Large, Dry-Type, Air Core Reactor Supported on Tall Concrete Columns.

FIGURE 5-142. A Synchronous Condenser Developed a Small Leak in Its Pressurized Nitrogen System.

ported on post insulators. Two types of rack failures have been seen with these racks. First, the welds on some racks were inadequate and the structural frames failed (Figure 5-143). In other racks of similar design, the post insulators supporting the rack failed (Figure 5-144).

The second type of capacitor bank, often found in conjunction with 500-kV systems, are massive structures. Some of these structures have special energy-dissipating devices to improve their earthquake performance (Figure 5-145). These structures have had limited seismic exposure. To date, none of these structures has collapsed, but the anchorage of components supported on the structures has failed and electrical connections between the racks and adjacent devices have been damaged as a result of inadequate flexibility.

A third type of capacitor bank is used with DC converter stations and takes two forms: suspended and frame structures. Suspended systems provide base isolation to the capacitors (Figure 5-146). They are restrained at the bottom to limit displacements from wind and earthquake-induced motion. One of these units experienced very high ground motions in the Northridge earthquake. While the capacitors were generally undamaged, the base restraints failed (Figure 5-147). This did not affect the capacitor's operation.

FIGURE 5-143. Capacitor Rack Welds Failed, Causing the Racks To Collapse.

FIGURE 5-144. Post Insulators Supporting Capacitor Bank Racks Failed.

FIGURE 5-145. Large Capacitor Racks Often Found in Conjunction with 500-kV Transmission Lines.

Note: In the lower left corner, a damping device supported on a concrete block can be seen behind the fence.

The frame structure style of capacitor bank is made of post insulators (Figure 5-148). Several post insulators in this structure failed, but the structure did not collapse (Figure 5-149). One would anticipate that, as some insulators failed, additional load would be imposed on the remaining insulators. This would lead to progressive collapse, which did not occur. The failed units may have failed at very low loads, as a result of the large variation in strength of porcelain. An alternative failure mechanism may be related to the fact that the length of post insulators can vary in length by about 1%. In an indeterminate structure, the short post insulators could have been subjected to large tensile stresses during assembly. The size of post insulators should be carefully matched or shims used to equalize their length.

A static var compensation facility can be a large facility rather than an item of equipment. The components within these facilities are similar to other items found in substations and will not be discussed further. None of these facilities are known to have experienced a significant earthquake.

FIGURE 5-146. Overview of Suspended Capacitor Bank.

5.16.1 Mitigation and Retrofit of Voltage Support Devices

As with other substation equipment, it is important to provide flexible connections that can accommodate relative motions between adjacent equipment.

5.16.2 Emergency Response Procedure for Voltage Support Devices

Some of these devices can be bypassed. Some of the capacitor banks used for 500-kV lines and for DC converter stations are needed for these facilities to continue to operate. Because of their importance for these high-energy-capacity facilities, special attention has been given to their seismic design.

5.16.3 Recommended Installation Practices for Voltage Support Devices

The seismic performance of capacitor racks has been mixed. The need to provide adequate conductor flexibility has been noted. The limited seismic exposure of the large earthquake-engineered capacitor banks leaves their relative performance largely untested. The suspended capacitor units expe-

FIGURE 5-147. Failed Restraints of Suspended Capacitor Bank.

rienced very high seismic inputs and suffered only minor damage to restraints. The cost of this approach compared to that of the frame-type capacitor rack is not known. It is important to avoid tensile stresses associated with the assembly of post insulators due to variation in their length. Indeterminate structures fabricated from post insulators, such as large capacitor racks, should use post insulators of the same length or shims to equalize their length.

5.17 STATION POWER

Station power is the AC power used directly by the control house and its equipment and indirectly by equipment in the switchyard for monitoring, control, communications, and operation of the substation. Within the

FIGURE 5-148. Overview of Frame-Type Capacitor Rack.

switchyard, there are typically two sources of station power, and some substations may have a feed from another substation.

Local station power is usually obtained from tertiary windings on a power transformer (Figure 5-150). The output typically ranges from 2 kV to 16 kV. The output may go to a bus supported on a structure and eventually to a station transformer. Although small three-phase transformers may be used, it is more common to have three pole- or platform-type transformers supported on the bus-support structure (Figure 5-151), on their own support structure, or on a concrete slab surrounded by a fence (Figure 5-152). These transformers are typically poorly anchored.

5.17.1 Earthquake Performance of Station Power

Station service has been disrupted by damage to the station power system. Clearly, if a station loses off-site power, station power will also be lost unless station batteries or standby generators are available and undamaged.

FIGURE 5-149. Damaged Post Insulator in Frame-Type Capacitor Rack.

Disruption has been caused by damage to tertiary bushings by interaction loads between the transformer and the bus support structure (Figure 5-150). While some unanchored slab-supported station transformers have moved, surprisingly, there is no record of this causing disruption.

5.17.2 Mitigation and Retrofit of Station Power

Although poor anchorage of the small transformers that provide station service has not yet caused problems, these transformers should be anchored. The lack of flexibility of connection to tertiary busses has caused bushing failures. These connections should be provided with adequate flexibility. Damage to the tertiary bushings can cause the entire transformer to be taken out of service.

FIGURE 5-150. Tertiary Bushings on This Transformer Were Damaged Because of Interaction Load from the Bus, Which Provides Power to Station Services.

5.17.3 Emergency Response Procedure for Station Power

Portable emergency generators have been brought to sites to provide temporary station service until normal station service could be restored.

5.17.4 Recommended Installation Practices for Station Power

Provide adequate flexibility to all station power connections. Anchor station transformers.

5.18 SUBSTATION CONTROL STRUCTURES AND THEIR CONTENTS

Substation control houses contain protective and auxiliary relays, communications equipment, control panels, station batteries, status indicator boards, and operator stations for personnel. The elements of a typical control house are identified in Figure 5-153. Not all control houses contain all of the components and systems shown.

5.18.1 Control House Structures

Most substations have a simple, one-story control house. In general, control house structures have performed well because of the small size and sim-

FIGURE 5-151. Station Transformer Supported on a Shelf That Is Part of the Bus Support Structure.

Note: The L-shaped sheet metal clip under the small radiator restrains the transformer but does not anchor it.

plicity of their structural systems. In California, they are often built using seismic provisions of building codes. Small reinforced masonry or prefabricated metal control house buildings have performed well, provided that the lateral-force resisting elements are adequately tied together. However, many are old and predate current building codes, or they are in regions that had no seismic requirements when they were built. Older control houses constructed with unreinforced or poorly reinforced masonry have suffered spectacular damage (Figure 5-154). Surprisingly, the substation and the equipment in this damaged control house continued to operate. Some control houses have adequate structural frames, but have unreinforced masonry infill walls.

While substation equipment is expected to age and will eventually require replacement, the control house can serve its purpose without modi-

FIGURE 5-152. One of the Most Common Methods of Supporting Station Transformers Is by Placing Them on Concrete Slabs at Grade without Anchorage.

fication. Because newer control equipment tends to be smaller than the equipment it replaces and many substations are becoming unstaffed, there is often no need to expand the control house, even though the substation may be expanding. Control houses should be seismically evaluated, and an appropriate upgrading schedule established. Because substations and their control houses are continuously in service, the use of shotcrete with reinforcing applied to the outside of the building is an option. The procedures for evaluating and seismically upgrading structures is beyond the scope of this document. However, there are several references issued by the Federal Emergency Management Agency (FEMA) dealing with the evaluation of existing buildings and seismic rehabilitation of buildings (see Section 8.1.1).

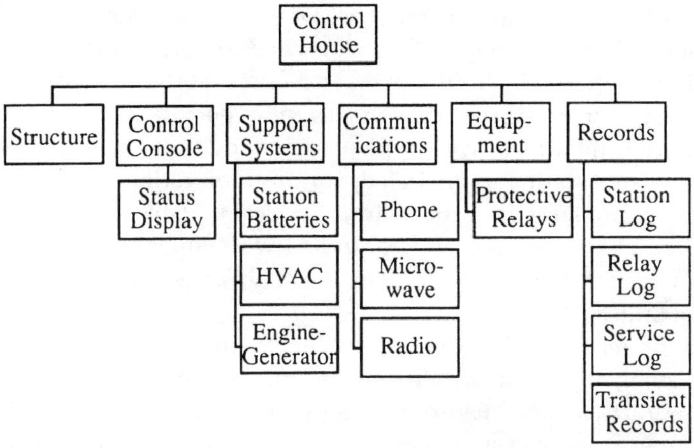

FIGURE 5-153. Diagram Showing Systems Contained in a Substation Control House.

FIGURE 5-154. This Unreinforced Masonry Control House Failed Spectacularly.

5.18.2 Equipment

A typical control house contains a broad range of equipment for monitoring and control, and several types of protective relays. These relays monitor conditions on transmission lines and in transformers. Many variables are monitored, including over- and under-voltage, voltage differential across transformers, the flow of current in circuits, direction of power flow, phase comparison, system frequency, and others. As substations are converted from staffed to unstaffed, various monitoring and communication devices are added, so the status of critical variables at unstaffed substations can be monitored and controlled with supervisory control from the staffed substations. These functions are typically performed by remote terminal units.

Some protective and auxiliary relays are vulnerable to chatter or changing of state in an earthquake. In general, although these actions may cause undesired actions and disruptions, they do not cause damage. Through action reports in nuclear power-generating plants, several specific relays have been identified as particularly vulnerable. For new construction, it may be useful to avoid these specific relays, especially since better performing alternatives are available at equivalent cost.

Substation instrumentation racks made in the United States and well anchored have performed well. Italian equipment observed in Chile had open backs and deformed in the earthquake.

5.18.3 Other Substation Equipment

Substations contain many other types of equipment, such as communications equipment, control panels, station batteries, status indicator boards, air conditioning, light fixtures, and computers. This equipment is also found in the energy control center and is discussed in Chapter 8, System Control, and Chapter 9, Communication Systems.

5.18.4 Recommended Practices for Substation Control Structures and Their Contents

Substation control houses should conform to seismic provisions in one of the major model building codes or to the local code, if that is more stringent. Some local jurisdictions may have reduced the requirement contained in the model codes, but this is not recommended for power facilities. Because control houses are small, simple structures, model code provisions should give good performance. Building codes are designed to prevent structural collapse; however, because substations must continue to operate after the earthquake, more stringent requirements may be appropriate.

5.19 MISCELLANEOUS FACILITIES: OIL STORAGE TANKS

Transformers and bulk oil circuit breakers contain large volumes of oil. Many substations use oil tanks to store oil while the units are being serviced. These tanks may be unanchored or poorly anchored (Figure 5-155). Flat-bottom tanks are found at some substations. The flat bottom tanks of the size typically found at substations have not exhibited the seismic vulnerability observed in larger tanks.

5.19.1 Earthquake Performance of Oil Storage Tanks

Some substation tanks have moved in earthquakes, but no tank failures have been known to occur. Tanks similar to those found at substations have moved and fractured pipe connections. A more complete description of the earthquake performance of oil storage tanks is given in Section 7.2.4.1

5.19.2 Recommended Practices for Oil Storage Tanks

It is recommended that flat-bottom tanks with a height-to-diameter ratio greater than 0.5 be anchored and that horizontal tanks be secured to their

FIGURE 5-155. Horizontal Tanks Are Frequently Unsecured to Their Cradles.

support cradles. Piping connections to tanks should be flexible to accommodate tank movement.

ENDNOTES

1. A.K. Chopra. 1980. *Dynamics of Structures: A Primer.* EERI.
2. Institute of Electrical and Electronic Engineers (IEEE). 1997. *Recommended Practices for Seismic Design of Substations.* Standard 693.
3. American Society of Civil Engineers (ASCE). Forthcoming. *Substation Structure Design Guide.*
4. A.J. Schiff, ed. 1995. *Northridge Earthquake: Lifeline Performance and Post-Earthquake Response.* American Society of Civil Engineers, TCLEE Monograph No. 8.

Chapter 6

TRANSMISSION AND DISTRIBUTION LINES AND SUPPORT STRUCTURES

Recommendations: Retrofitting and New Construction

The recommended practices for distribution transformers are contained in Section 5.8.

Recommendations: Retrofitting

Materials for a transmission tower should be kept in stock, so that a damaged tower can be replaced without having to wait for a unit to be manufactured by a steel fabricator (6.1.3).

Recommendations: New Construction

When possible, transmission line towers should be positioned back from steep slopes (6.1.4).

The design of transmission tower foundations located on ridges should take ridge shattering into consideration (6.1.4).

The design of transmission tower foundations near water, such as at river crossings, should consider liquefaction and lateral spreading (6.1.4).

6.1 TRANSMISSION SYSTEMS AND THEIR SUPPORT STRUCTURES

Transmission lines connect substations and generating stations in the power network and carry power at transmission voltages. Some utilities also have lines that are referred to as *subtransmission lines*. Often, these lines served as the transmission system before a new network operating at a higher voltage was superimposed on the old system. Some utilities have adopted different design and construction practices for their transmission and subtransmission systems. In this document, no distinction will be made between transmission and subtransmission. The transmission voltages vary between utilities. Utilities that serve less-populated areas may have transmission voltages that can be found in the distribution system of larger utilities. Transmission lines are typically supported by steel structures, although wood poles or structures are also used. On rare instances, transmission lines are buried.

6.1.1 Earthquake Performance of Transmission Systems and Support Structures

The performance of transmission lines and their towers has generally been good. In the 1992 Landers, California, earthquake, a fault passed under a tower with two legs on each side of the fault. The horizontal fault offset at that location was 3 m (9.5 ft) (Figure 6-1). Although the tower was deformed, it continued to support the lines.

In the 1995 Kobe, Japan, earthquake, there was damage to post insulators on dead-end structures. The post insulators were used to restrain the transmission line as it looped under the crossarm supporting the in-line insulator strings. The insulators broke at the base of the insulator connected to the tower and ended up being supported by the line. This did not disrupt service, although the insulators had to be replaced.

In the Northridge, California, earthquake, several towers were damaged. Damage was primarily attributed to foundation failures. Towers on ridges and crests of hills had foundations fail as a result of ridge shattering (Figure 6-2). In one case, when a tower failed, the two towers on each side were also pulled down by the lines. The dead-end structure discussed in Chapter 5, Substations, was similar to a transmission tower, and its failure was attributed to soil liquefaction.

While soil liquefaction may pose a significant risk to transmission towers in some parts of the country, no failures during U. S. or foreign earthquakes have been observed, except as noted above.

There have been cases where landslides and avalanches damaged transmission towers. These events were not triggered by earthquakes, but, clearly, earthquakes could trigger such events.

The liquefaction and lateral spreading adjacent to rivers and other bodies of water suggests that towers in these locations, such as at a river crossing, may have higher vulnerability.

6.1.2 Mitigation and Retrofit of Transmission Systems and Support Structures

The earthquake performance of transmission towers has been so good that no recommendations for mitigation or retrofitting are made. Some utilities, however, keep spare towers or emergency towers in storage so that, should a tower fail, a replacement can be installed without waiting for suppliers to fabricate a new tower.

FIGURE 6-1. A Fault with an Offset of More Than 3 m (9.5 ft) Passed between the Legs of a Transmission Tower But the Tower Did Not Collapse.

6.1.3 Emergency Response Procedure for Transmission Systems and Support Structures

There have been cases where it was desired to quickly put a transmission line in service without waiting to construct a new tower. In one case, a large crane was used as a temporary transmission tower (Figure 6-3). To the right of the crane is a prefabricated emergency transmission tower. This unit is commercially available and could be held in reserve to quickly replace a tower that is damaged.

6.1.4 Recommended Installation Practices for Transmission Systems and Support Structures

It is advisable to carefully evaluate the foundations of transmission line towers that are to be placed on steep ridges to account for ridge shattering.

FIGURE 6-2. Transmission Tower Collapse Caused by Ridge Shattering and Foundation Failure.

When possible, transmission line towers should be set back from the edge of steep slopes. Because of the lack of performance data on transmission towers in liquefiable soils, no recommendations are given. However, foundation designs for towers near water, such as at river crossings, should consider liquefaction and lateral spreading.

6.2 DISTRIBUTION SYSTEMS AND SUPPORT STRUCTURES

The distribution system, which typically operates between 4 kV and 34 kV, encompasses the lines and equipment beyond the distribution substation that carry power to customers. The distribution system typically consists of radial feeder lines, poles, reclosers, manually operated switches, remotely operated switches, fuses, capacitors, and transformers that are pole-elevated or platform-, pad-, or vault-mounted. In the United States, aerial distribution lines are primarily supported on wooden poles. Installations in newer residential areas now typically use buried lines and pad- or vault-mounted distribution transformers. Outside of North America, concrete poles are usually used to support distribution lines. Reclosers are small circuit breakers that open when a line is overloaded but automatically try to reclose several times before locking open. While the distribution system is radial in character, there may be some ties or switchable links between systems from different substations. Some urban areas are configured as a grid

FIGURE 6-3. A Construction Crane Was Used To Temporarily Support a Transmission Line To Get It Back in Service.

to improve system reliability. The transformers that provide power to the customers were discussed in the Section 5.7, Power Transformers.

6.2.1 Earthquake Performance of Distribution Systems and Support Structures

Two types of failures to the distribution system, excluding distribution transformers, have been common: burn-down of feeder and service lines and failure of concrete distribution poles. While less common, soil liquefaction can cause unanticipated failures.

Feeder lines can be set in motion by the earthquake, and the vibration of poles can cause changes in tension in the lines so that lines come in contact. When energized lines come in contact, reclosers may operate, fuses may blow, or lines may burn through or wrap around each other. The burnt-through lines can fall to the ground. Downed lines can remain energized, so they pose a serious safety hazard. They have started fires, disrupted traffic when they fall across roadways, and created a large load on the emergency response community. In recent California earthquakes, a large percentage of emergency response calls were associated with downed power lines, and a large percentage of the earthquake-related fires were caused by downed lines. In some areas, lines of different voltages will be carried on the same pole. There have been problems of an upper line falling on a lower voltage line. There have been many reports of pole-mounted transformers blowing up or catching on fire as a result of the earthquake and the vibratory response of the lines. It appears that, if the flow of power is disrupted in the area before local shaking builds up, there will be much less disruption to the distribution system, as lines will not burn down, fuses will not be blown, and transformers will not be damaged.

In the United States, wood distribution poles are typically used and their performance has been very good, although there have been some failures as a result of soil failure or dry rot in the poles. One utility had plans to replace 600 dry-rot-damaged poles. In some poles, only the outer inch had not deteriorated. An earthquake occurred before the replacement project started. Surprisingly, there were no pole failures in the service area.

In some countries, it is common to support distribution lines from existing buildings; this often happens in old communities where there is no room in the narrow streets to install poles. Thus, lines are often lost when the structure fails. In Japan, hollow concrete poles are used extensively. In the Kobe earthquake, many poles appeared to fail as a result of vibration (Figure 6-4). In areas of liquefaction, poles sank into the ground; in one case, the pole settled more than 5 ft (Figure 6-5).

The overall impact of earthquake damage to the distribution system in California has been characterized as being equivalent to that caused by a severe winter storm.

6.2.2 Mitigation and Retrofit of Distribution Systems and Support Structures

While there has been significant damage to distribution systems, mitigation and retrofitting does not appear justified, with the possible exception of anchoring or restraining distribution transformers.

FIGURE 6-4. During the Kobe, Japan, Earthquake, Concrete Poles Were Severely Damaged.

6.2.3 Emergency Response Procedure for Distribution Systems and Support Structures

Two potential problems related to emergency operations have been observed in past disasters. First, if there are lengthy power disruptions, homeowners have sometimes hooked up small emergency generators to provide basic power needs. There have been cases where the homeowners have hooked their generators to the house wiring system without opening the main power disconnect. As a result, the distribution system was energized by the homeowner, and utility personnel have been electrocuted. Grounding lines that are not to be worked on hot is a good precaution.

FIGURE 6-5. This Pole Sank More Than 5 ft When the Soil Liquefied.

When the power system is re-energized prematurely, there is the potential that downed lines will become a source for fire ignition. After a damaging earthquake, it is important that re-energizing the distribution system be done in coordination with the local gas companies and fire services.

6.2.4 Recommended Installation Practices for Distribution Systems and Support Structures

The only observation from past earthquakes to improve the performance of the distribution system is to anchor distribution transformers as they are installed. The recommended practices for distribution transformers are contained in Section 5.8.

Chapter 7

POWER-GENERATING FACILITIES

Recommendations

Combustion-turbine units used for black-start at steam-generating stations should have all support systems supplied with emergency backup power, and their control systems should be able to operate without a frequency-synchronizing signal from the power grid (7.1).

Steam-generating station sites located adjacent to bodies of water may be vulnerable to soil liquefaction. Designs should take into account the potential for differential settlement and lateral spreading, which can affect turbines, liquid storage tanks, and stacks (7.2.1.5, 7.2.4.1, and 7.2.4.2).

The gap between the turbine pedestal and the turbine building operating floor should be large enough to accommodate relative motion between the structures and to avoid impacts (7.2.1.2).

Reliable emergency backup power should be provided to operate turbine main-bearing lubricating oil pumps and turning gear (7.2.1.3).

The use of high-pass filters should be considered on vibration sensors used to monitor turbine unbalance (7.2.1.4).

Piping and other interfaces with the steam generator must be able to accommodate earthquake-induced movements between the steam generator and its support structure (7.2.2.2).

Piping that spans different structures must be designed to accommodate earthquake-induced relative motions (7.2.2.3).

Small pipes, such as measuring taps, that are connected to larger pipes must be provided with adequate flexibility to accommodate the earthquake-induced movement of the large pipe (7.2.2.3).

Cabinets, such as motor control centers, relay racks, control consoles, and switchgear, must be adequately anchored (7.2.4).

See Sections 8.1.2 and 8.2 for seismic issues related to the control room.

Coal silos and conveyor systems should be designed for seismic loads (7.2).

Liquid storage tanks should be designed for seismic loads. Preventive maintenance should address corrosion of the weld between the sidewall and base of the tank. The use of J-type anchor bolts to anchor the tank is not recommended. (7.2.4.1)

Piping connections to tanks should be flexible, so rocking of the tank does not cause pipe failures (7.2.4.1).

Horizontal tanks should be secured to their cradles, and lateral restraint should be provided to tanks with expansion roller or slide plate supports (7.2.4.1).

Circulating water system pipes that transport water between plant heat exchangers and cooling towers, rivers, or reservoirs, should be designed to accommodate movements associated with lateral spreading (7.2.4.3).

Preventive maintenance is needed on forced-draft cooling towers, so they retain their structural integrity (7.2.4.4).

For at least one elevator that services the steam-generator building, earthquake provisions of the current elevator code should be reviewed to determine whether cost-effective retrofits can be made to improve its earthquake performance (7.2.4.5).

In this chapter, the earthquake performance of power-generating stations will be reviewed and, because of their generally good performance, recommendations will be made only for the specific areas that have performed poorly in the past. The standard practices that have performed well will not be discussed.

Power-generating stations usually have step-up transformers and high-voltage switchyards at the terminus of the transmission line that connects the generating unit to the power grid. The design and installation of these components are the same as used in substations and reference should be made to Chapter 5.

Power-generating facilities of two types are considered here: combustion-turbine and steam-turbine generating units. Hydropower and diesel power are not explicitly discussed, but much of the equipment associated with those generating facilities is discussed in this document. Solar generating facilities are not discussed because of their limited use. This discussion excludes nuclear generating plants since they are under federal regulation and are designed to much more stringent criteria. The material on switchyards in Chapter 5, Substations, is applicable to switchyards at nuclear generating facilities.

7.1 COMBUSTION-TURBINE GENERATING UNITS

Combustion-turbine generating units, fueled by gas or oil, are typically used for peaking power but may also be used for base load. They are sometimes located at steam-turbine generating stations. Peaking units typically have capacities of 25 megawatts (MW) to 50 MW. Small units are often trailer-mounted on jacks. In recent years, there has been an increased use of large combustion turbines in combined-cycle generating stations.

Combustion-turbine generators are inherently rugged, and no seismic damage has been reported, although their seismic exposure has been limited. Damage to ancillary systems and equipment such as fuel tanks and control panels could cause lengthy disruptions. When units are located at a steam-generating station, one combustion turbine may be designed to provide black-start power in the event of a total loss of power at the station. There have been several earthquakes where the combustion turbines failed to perform their black-start function. Failures appeared to be related to minor problems not directly related to earthquake damage. In one case, a critical component was powered by station power, which in a blackout would not operate. In other cases, the control system was dependent on an AC signal for synchronizing the unit to the power-grid frequency. Without this signal, which was absent in a blackout condition, the unit would not start.

7.2 STEAM-TURBINE GENERATING UNITS

A description of the operation and components of fossil-fueled steam-generating stations will not be attempted here. However, their earthquake response has been good to date. Observed damage and suspected damage have not been related to the system operation or configuration, except as noted below. Large coal-fueled plants have not been subjected to damaging earthquakes, and even combustion gas- and oil-fueled units have seen only limited exposure. Most of the latter plants have been relatively small, generating less than 40 MW.

Although all types of fossil-fueled steam-generating units have similar systems for each function, from a seismic point of view fossil-fueled steam-generating units can be divided into two categories: oil- and gas-fueled units and coal-fueled units. Because of the combustion process, a coal-fueled steam-generator is physically larger than an equivalent capacity gas- or oil-fueled steam generator. Also, coal-fueled units have massive coal storage silos high in the boiler structure, which may increase their seismic vulnerability. They also have extensive coal handling equipment, which is often designed without consideration of seismic loads.

Steam-turbine generation stations have a control room from which the operation of the station is monitored and controlled. Seismic issues relevant to the station control room are discussed in Section 8.1–8.3, 8.5.1, and 8.5.3.

Damage to steam-turbine generating units can be grouped into five categories of equipment or facilities: turbines, steam generators, commercially produced equipment, engineered equipment, and structures.

7.2.1 Turbines

Earthquakes have caused several different types of damage to turbines and caused systems to go off line.

7.2.1.1 Damaged Thrust Bearings. Earthquake motions along the axis of the turbine can load and exceed the capacity of the thrust bearings, resulting in their being wiped. That is, there is metal-to-metal contact that damages the bearing surface. This will require early maintenance, but the turbine can usually continue to operate for some time. If the axial motion of the rotor relative to the housing is large, the moving blades will come in contact with the fixed blades and cause major damage that will result in very long repair times. This is apparently an unavoidable design feature of the turbine. A small steam turbine was damaged in this way in the 1972 Managua, Nicaragua, earthquake (Figure 7-1). In the United States, the wiping of thrust bearings was reported in an earthquake with a Richter Magnitude of only 4.2.

7.2.1.2 Impact Between the Turbine Pedestal and the Operating Floor of the Turbine Building. The turbine is typically supported on a massive, stiff turbine pedestal, which has its own foundation. At the turbine floor level in the turbine building, which is generally about two floors (9–12 m) above its foundation, an isolation gap usually separates the operating floor of the turbine building from the turbine pedestal to keep the turbine-generator vibrations from being transmitted to the surrounding building. If the gap is not large enough, there can be hammering between the pedestal and the operating floor when the structures are out of phase during earthquake motion. While local spalling at contact points has been observed, the more significant damage from impacting can be to the thrust bearings of the turbine. Also, if there is sufficient longitudinal motion of the rotor, fixed and moving blades can come in contact, causing major damage to the turbine.

7.2.1.3 Loss of Emergency Power. Turbine bearings have also been damaged as a result of the loss of off-site power, typically from damage to the switch-

FIGURE 7-1. A Small Steam Turbine Was Severely Damaged When Axial Loads Caused the Moving and Fixed Blades To Interact.

yard, and the simultaneous failure of station power (batteries and/or emergency generators). If damage to the switchyard prevents turbine power from being exported, the generator will trip off line and the turbine will coast to a stop. Without station power, lubricating oil pumps will not operate at low turbine speed, and the bearings can be damaged. Also, without station power, the turning gear does not operate, and the hot turbine-rotor shaft can sag and take a permanent set as the turbine cools.

There has been one case where all power in the generating station was lost. Station control, lights, communication equipment, base station radio for the station, and elevators were all dead. The only means of communication was handheld radios, and the only sources of light were battery-operated flashlights. Station power to the plant was normally provided either by transformers in the switchyard or by a small house turbine. At the time of the earthquake, the house turbine was down for maintenance and the transformers were off line because of damage in the switchyard. Power to drive the lubricating pumps was lost, and the main bearings of the turbine generator had to be replaced. After the earthquake, the utility installed an emergency engine-generator at the plant.

7.2.1.4 Triggered Vibration Sensors. Large turbines typically have several vibration-monitoring sensors to indicate whether the turbine rotor is out of balance or to shut the system down if unbalance exceeds a specified level. In Japan, there have been several instances where vibration sensors mounted on the turbine's external housing have tripped the turbine off line. When a unit goes off line, it frequently takes from several hours to a full day to get it back on line. To eliminate unnecessary tripping of the turbine, some Japanese utilities have placed a high pass filter in the vibration transducer signal path before the signal goes to the level monitor. The sensor is intended to detect turbine unbalance, which is at a frequency much higher than earthquake vibrations. The filter eliminates the low-frequency earthquake signal and retains the unbalance signal.

7.2.1.5 Differential Settlement of Steam-Turbines and Their Piping. Power plants located adjacent to rivers or bodies of water may be vulnerable to soil liquefaction, although this type of damage has not yet been observed in the United States. Small differential settlements of the turbine could cause severe problems. In the 1995 Kobe, Japan, earthquake, a generating plant did experience differential settlement, and the plant was out of service for months.

7.2.2 Steam Generators and Support Systems

7.2.2.1 Steam Generator Internal Damage. In the 1978 Sendai, Japan, earthquake, an internal bracket in the steam generator (boiler) of a large oil-fueled generating unit failed. It then acted as a battering ram to damage the pipes that it was meant to restrain. This type of damage was indicated by excessive steam in the stack output. Internal water leaks could also be observed by a drop in water pressure or the excessive use of make-up water. Although repairs were simple, it took almost a week for the boiler to cool down so that personnel could access the damaged area. Minor boiler damage in a steam generator also occurred in the Loma Prieta earthquake.

7.2.2.2 Steam Generator Support Structure–Steam Generator Interaction. Most large plants have steam generators suspended from the top of the steam generator support structure. This design accommodates large vertical deflections associated with thermal expansion as the steam generator heats up and cools down. The steam generator support structure (that is, the building) can be closed (have exterior walls) or open. On open structures, the steam generator requires restraints between the steam generator and the support structure to limit wind-induced motion. However, seismic restraints are needed for both open and closed structures. The restraint systems vary, depending on the design of the steam generator manufacturer. The steam generator is surrounded by trusses or girders at various levels, which serve to resist internal pressures within the steam generator. These trusses or girders, often referred to as *buckstays*, also serve as a means for providing lateral restraint to the steam generator. Bumpers or hydraulic ties between the buckstays and the steam generator support structure provide the seismic restraint.

The ties that restrain a steam generator for wind loads are often inadequate for seismic loads and may fail in an earthquake. Restraints are often referred to as *seismic stops*, or bumpers, and may be sacrificial. That is, they may sustain severe damage but still perform their functions (Figure 7-2). They are easily repaired, and they protect the steam generator from sustaining severe damage.

The various piping connections and other interfaces between the steam generator and its support structure are designed to accommodate the thermal growth movement. However, they may not be able to accommodate the lateral movement associated with earthquakes. Examples include highly restrained piping (long runs of pipe will be flexible enough to accommodate

FIGURE 7-2. A Seismic Stop.

Note: Seismic stops are provided to limit lateral movement of a suspended steam generator. These stops may be damaged but still perform their function.

relative motions) and duct work. Issues related to piping are discussed below. There has been damage to ductwork between the main steam generator air supply fan and the steam generator as a result of movement of the steam generator.

7.2.2.3 ***Piping.*** In general, welded steel pipe is very rugged. Indeed, long runs have been observed to experience large displacements and strike structural members with no damage except to the thermal insulation (Figure 7-3). There has also been damage to the packing between piping and the steam generator shell, where pipes enter the steam generator. This damage is superficial, does not affect the operation of the plant, and is easily repaired. Some damage has occurred to small inflexible pipes that are attached to larger, long pipe runs that are more flexible. An example is a small tap for instrumentation connected to a major steam pipe. Threaded pipe connections have frequently failed. Damage has been observed where pipe runs penetrate walls separating different structures so that relative motion between the structures serves as a guillotine for the pipe. Similarly, pipes that are anchored to different structures that experience relative deflections are frequently damaged. Piping running between or connected to different structures must be provided with adequate clearances and flexibility to accommodate the relative motion of the structures.

FIGURE 7-3. Restraints to a Large Steam Pipe Were Severely Deformed as a Result of the Dynamic Response of the Pipe Relative to the Restraints.

7.2.2.4 Fuel Supply. For gas-fueled units, there may be a disruption in the gas supply, which is usually provided by a gas distribution line. Historically, the earthquake performance of gas transmission lines has been very good. However, lines crossing faults or rivers are vulnerable. The risks to the fuel supply of oil-fueled plants are similar to those of gas-fueled plants. Oil-fueled plants usually have large fuel-oil storage tanks on site. The seismic performance of tanks is discussed below.

For coal-fueled units, there may be problems with coal supply. Today, most plants store only one to two months' supply of coal at plant sites. The coal supply may come by river barges or on special unit trains that travel over distances of more than a thousand miles. An example of a long-term disruption not associated with an earthquake occurred in the 1993 Midwest flood. This flood severely disrupted the fuel supply to several coal-fueled generation plants. The time for unit trains to deliver coal more than doubled as a result of the loss of direct routes and the need to travel slowly because of road-bed damage and congestion. Deliveries were also limited by the lack of engines, coal hopper cars, and crews. In the fall of 1997, without problems caused by a natural disaster, shortages of engines, crews, and cars caused a disruption of rail shipments that threatened to shut down generating stations. In the 1886 Charleston earthquake, long sections of the rail system were destroyed.

Coal-fueled plants have not been subject to significant earthquakes, so their seismic performance is unknown. A seismic review of some of their facilities suggests that coal-handling conveyor systems may be vulnerable. For example, in one plant, a conveyor supplies coal to the plant silos at about the 80 m elevation of the plant. The structural connection for the conveyer support structure is a pin at the bottom and a roller at the top. The top roller rests on a 15-cm-deep platform. In an earthquake, if the steam generator support structure at the top of the conveyer moved more than 15 cm, the conveyer could fall from its upper support and the plant would be out of operation for a significant period.

7.2.3 Commercially Produced Equipment

The term *commercially produced equipment* refers to equipment that is produced according to a manufacturer's standard design. Examples include pumps, motors, motor control centers, and low- and medium-voltage switchgear. This type of equipment has generally performed very well in earthquakes. Most problems have been associated with inadequate anchorage. The anchorage on some equipment, such as pumps and motors, that has relatively high service loads, performs well in earthquakes since normal service loads exceed earthquake loads. Smaller equipment that is shipped assembled usually performs well, since the shipping loads can exceed earthquake loads.

7.2.4 Engineered Equipment

The term *engineered equipment* refers to equipment that is designed on a plant-specific basis. Examples of this include coal-handling equipment, cable trays, piping systems, tanks of various types, and large fans. In general, this type of equipment has performed well. Standard industrial practice has yielded very good seismic performance of cable trays and piping. Coal-handling equipment, discussed above, may be more vulnerable. Damage to engineered equipment is often associated with relative deflections among different structures that the equipment spans.

7.2.4.1 Liquid Storage Tanks. Power-generating stations typically have many different types of liquid storage tanks. Large liquid storage tanks have exhibited several failure modes in past earthquakes. Some of the observed failures were at facilities other than generating plants. Unanchored or inadequately anchored flat-bottom tanks have rocked, causing rigid pipe connections to fail.

Rocking has also caused "elephant-foot" buckling—that is, a bulge around the base of the tank. Elephant-foot buckling (usually in conjunction with corrosion) or corrosion of the weld connecting the side wall to the base plate of the tank can cause relatively large ruptures (Figure 7-4). The consequent rapid emptying of the tank can generate a low internal pressure in the tank, causing it to buckle inward at the top (Figure 7-5). When there is elephant-foot buckling, the walls of tanks have torn at stiffeners surrounding access ports.

FIGURE 7-4. The Corroded Weld Joining the Side Wall to the Base Plate Failed, Causing a Loss of Contents.

FIGURE 7-5. Several Fuel-Oil Storage Tanks Failed, and the Rapid Loss of Contents Caused the Tank To Buckle.

J-type anchor bolts have pulled from their embedments, allowing the tank to rock, which has contributed to elephant-foot buckling. Sloshing of fluid in a tank that is relatively full can also deform the top and upper walls of the tank. Sloshing of contents can damage lids on floating top tanks and damage internal roof supports. Smaller tanks can slide, so that rigid pipe connections fail or drain pipes that extend below the tank are sheared off. Rigid connections between tanks, such as catwalks or piping, may fail due to relative motion of the tanks. Soil below tanks has liquefied, and lateral spreading has damaged pipe connections.

Horizontal tanks supported in cradles have shifted because they were not anchored to their cradles. Plant feedwater heaters are usually supported on wheels on one end to accommodate thermal growth of the tank. These wheels need side restraints for seismic protection. Smaller vertical tanks supported on legs have fallen over when legs buckle.

7.2.4.2 Stacks. There are records of damage to stacks (chimneys) dating back to the 1886 Charleston earthquake. Most damage has been to brick stacks and appears to occur between two-thirds and three-quarters of the height. Power plant stacks are constructed with an outer reinforced concrete shell and an internal liner, which may be free-standing or supported by the shell. In the Loma Prieta earthquake, the stack liner at a generating station broke away from the outer shell by which it was supported. Repairs included adding straps to the outer shell (Figure 7-6).

FIGURE 7-6. Straps Were Added to the Stack after the Loma Prieta Earthquake, in Which the Connections between the Liner and the Shell Were Damaged.

At one plant that had water storage tanks in the base of the stack, piping connections to the tanks failed. The tank did not appear to rock within the stack, so rocking of the stack may have damaged the pipe connections.

7.2.4.3 Cooling Water Systems. Generating plants are often located adjacent to large bodies of water, which are used as a source of water to condense the steam back to water once it has expended its energy in driving the turbine. In this once-through cooling system, there has been damage to the large circulating water pipes that transfer water between the generating unit condenser and the water source as a result of the earthquake-induced movement of pipe supports in liquefied soil at or near the shoreline.

7.2.4.4 Cooling Towers. When a reservoir cannot be used, power stations use either natural draft (hyperbolic-shaped) cooling towers or mechanical draft (forced-air) cooling towers. There has been damage to forced-air type cooling towers at a non-power station facility that used a design similar to those found at power stations. Failure was attributed to structural members that appeared to have been weakened by corrosion. Natural draft cooling towers have not experienced earthquake damage.

7.2.4.5 Elevators. Elevators in power-generating stations are more than a convenience. Generating stations 60 m high are common, and units as tall as 90 m can be found. If plant repairs are required, moving tools such as welding equipment and repair parts to the height equivalent of a 20-story building without elevators is not practical.

The earthquake performance of elevators has been marginal. Seismic provisions for elevators have only been mandatory requirements of the national elevator code for a few years, so most power plant elevators do not comply with the good practices now required in elevator codes. In general, the performance of freight elevators has been better than for passenger elevators. Simple retrofit procedures to improve earthquake response should be considered so that access in the steam generator structure is assured. Things to consider include anchoring control cabinets, arching the hoist beam, and making improvements to reduce the chances of cables being fouled.

7.2.5 Structural Damage

Structural damage to power plants has been very limited. Isolated members have buckled and some connections have failed (Figure 7-7), but the

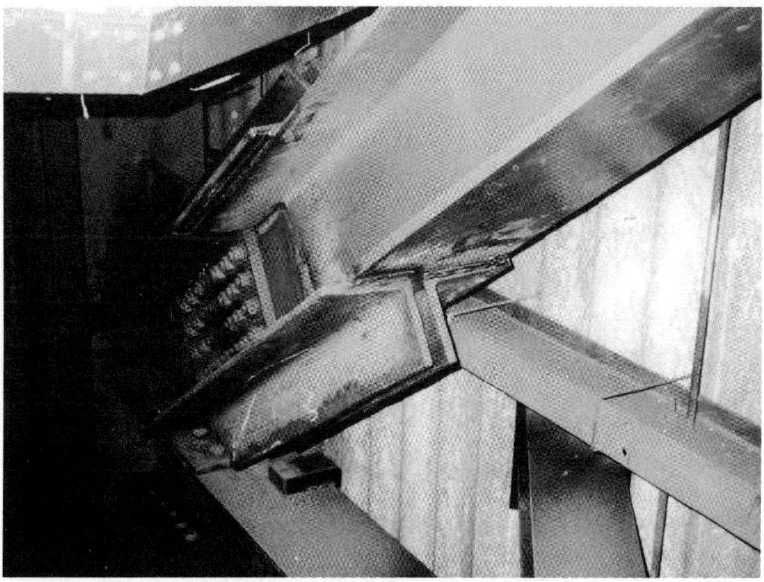

FIGURE 7-7. A Brace in the Steam Generator Support Structure Failed and Is Shown after It Was Repaired.

operation of the plant was not affected and damaged members were easily repaired.

Typically, one wall of the turbine building is shared with the steam generator support structure and the opposite wall is free-standing. The typical high bay of the turbine building means that the free-standing wall will be relatively flexible perpendicular to its length and may be damaged. The outside wall also typically supports one of the rails for the overhead crane needed to service the turbines. When not in use, the crane should be parked at one end of the turbine building. If the crane is located away from its end position, the seismic movement of the outside wall of the turbine building may damage the rails or wheels on the crane. At worst, the crane could fall from the support girder.

Chapter 8
SYSTEM CONTROL

Recommendations: Retrofit and New Construction

Suspended control room ceilings should have restraints to reduce lateral movement (8.1.2).

Light fixtures on suspended ceilings should also be restrained from a structural member to prevent them from falling (8.1.2).

Stem-style suspended light fixtures should have safety restraints (8.1.2).

Air diffusers should have attachments to ducts that do not depend on friction (8.1.2).

Water pipes and appliances, including sprinkler systems in and above the control room or control room equipment, should be supported or restrained to withstand seismic loads (8.1.2).

Control center support systems, such as HVAC, lighting, computers, control equipment, and communication equipment, should be evaluated to determine if they need UPS or emergency backup power (8.2.2, 8.3.13).

The structural integrity and installation of UPS and backup power systems should be evaluated and seismically upgraded (8.2.2.2, 8.2.2.3, 8.2.2.4).

Installation of engine-generator systems, including the engine-generator, control console, starting system, day tank, main fuel tank, piping system, oil cooler, cooling system, exhaust system, and transfer switch, should be evaluated and seismically upgraded (8.3.1through 8.3.10).

A schedule for periodic testing of emergency power systems should be established and followed (8.3.11).

Operating procedures for emergency power systems should be documented and posted near the systems (8.3.12).

The installation of an external power hookup for a mobile generator should be considered (8.3.14).

Utilization of mutual aid and government emergency support systems should be incorporated into emergency management systems (8.4.1, 8.5.4.5).

The structural integrity and installation of control consoles, status boards, SCADA systems, and computer systems should be evaluated and seismically upgraded (8.5.1, 8.5.2, 8.5.3).

Raised floors used in computer rooms should have floor-support pedestals bolted to the subfloor or held with high-strength bonding materials (8.5.3.2).

Equipment on computer floors should be restrained from lateral movement and from tipping over (8.5.3.2).

Recommendations: Retrofit

The mitigation program should include the eventual seismic renovation or replacement of vulnerable buildings housing the control center and other critical control functions (8.1).

Recommendations: New Construction

The design of structures containing control centers or other critical control functions should address serviceability issues as well as meet seismic requirements of building codes (8.1.1).

Raised computer floors with stringers anchored to pedestals perform best (8.5.3.2).

System control involves the control of power output from generating stations to keep the supply of power balanced with the demand. This is often referred to as *power dispatch.* It also involves the control of the configuration of the power network so that damaged facilities can be isolated and/or bypassed to maintain service and to clear circuits for repair. Issues related to system control also include identification of system damage, restoration of service, and the repair or replacement of damaged equipment.

Issues related to system control are discussed from the perspective of the utility's control center, although similar equipment can also be found in substations. Communications, which are an important part of the control center, are discussed in Chapter 9. Equipment commonly found in both control centers and substations includes: uninterruptible power supplies (UPS), control consoles and boards, display boards, and the supervisory control and data acquisition (SCADA) system. The use of SCADA systems is increasing as more substations become unstaffed. In substations, these systems monitor various system variables and the status of the power system equipment and then transfer the data to the control center. Equipment generally found only in the control center includes: emergency power engine-generator(s), computers, and large HVAC systems.

Many vital functions necessary for the operation of a utility's power system are performed in the control center. A schematic diagram for a typical control center is shown in Figure 8-1. The control center has been divided into four groups of equipment, systems and functions: the structure, support systems, system control, and the facilities related to damage identification and restoration (damage restoration).

8.1 CONTROL CENTER STRUCTURE

The control center structure contains most of the control system that is not distributed to other facilities. Some utilities may have these functions located in more than one structure.

Few such facilities have been subjected to significant earthquakes, and none have been damaged or significantly disrupted by the effects of an earthquake. At one control center, located on the fifteenth floor of a building, the loss of power to the building required personnel to climb stairs to get to the control center. The control center itself was provided with emergency power.

8.1.1 Structural Systems

The seismic vulnerability of the control center structure should be reviewed. This review should include the evaluation of the vulnerability of

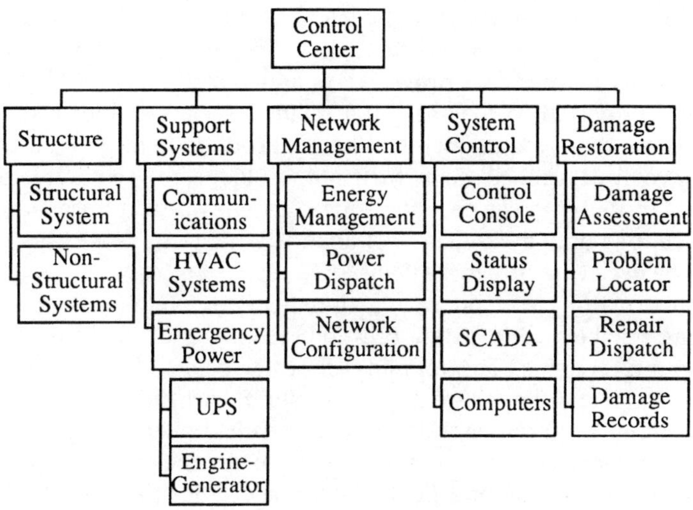

FIGURE 8-1. Schematic Diagram of Control Center Functions and Systems.

the site, its accessibility, and the structural integrity of the building. Since the control center must remain operational after an earthquake, the normal design criteria used in building codes may not be appropriate. Building code criteria require only that a building not collapse when subjected to a major earthquake. While the building codes do have increased design criteria for buildings that house essential functions and while the codes penalize elements with poor ductility by increasing design coefficients, they do not explicitly address serviceability issues. The evaluation and mitigation of damage to structures is beyond the scope of this document. However, there are several references issued by the Federal Emergency Management Agency (FEMA) dealing with the evaluation of existing buildings[1–3] and seismic rehabilitation of buildings.[4–10]

It is recommended that seismic upgrading of inadequate structures housing control systems be incorporated into the mitigation plan.

8.1.2 Nonstructural Systems

In addition to the building's structural system, there are nonstructural and support systems. The HVAC systems and emergency power system (UPS and engine-generator systems) are discussed below. There are other building components that can cause injury or damage equipment should they fail.

Control rooms often have suspended ceilings, and their earthquake performance, even when they meet code requirements, has been marginal. Figure 8-2 shows a collapsed ceiling resting on a control console. This type of failure has the potential for injuring personnel and disrupting system control at an especially critical time—the immediate postearthquake period. T-bar-supported ceilings often drop ceiling panels, light fixtures, and air diffusers. Most ceiling panels are lightweight, and their dropping is an inconvenience rather than a serious hazard; however, some panels are heavy and pose a hazard. Seismic code provisions require that light fixtures have safety supports connected to the structural ceiling; however, older installations may predate these requirements. While the code requires that suspension wires have three turns to secure their support loops, this requirement is frequently ignored in areas of low seismic awareness. Air diffusers are often simply clipped onto ducts. It is recommended that positive anchors, such as screws, be used to secure these items.

Most structures contain several independent water piping systems, including potable water, fire suppression systems, and chilled water for air conditioning systems. The toppling of water cooler drinking fountains, water heaters, and the failure of pipes or sprinkler heads can cause significant flooding. Even minor water leaks can spread over a floor and leak to lower floors far removed from the original leak. In one case, an unanchored water heater on the second floor at a service center toppled in an earthquake, and, as a result, water penetrated to the communication room below and disrupted all communications to the service center.

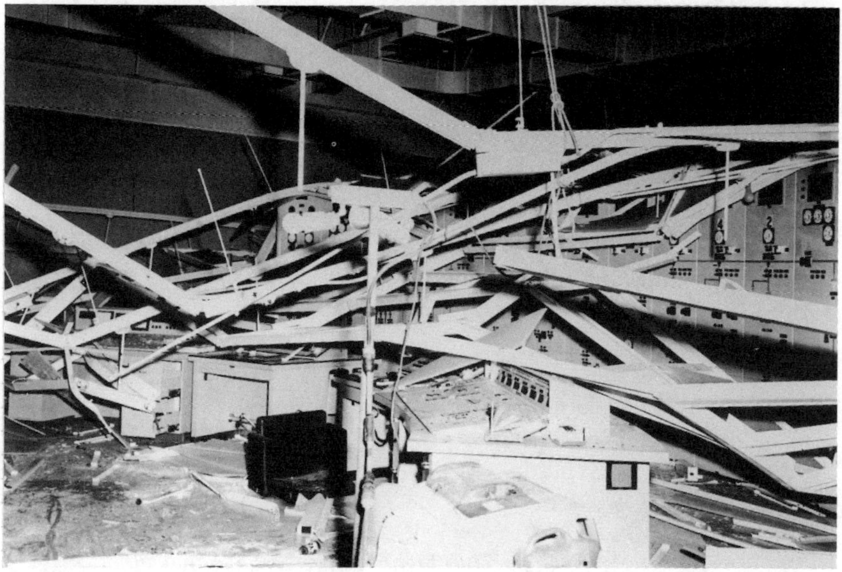

FIGURE 8-2. Collapsed Ceiling in a Control Room Can Make It Impossible To Control the System.

It is recommended that all appliances with water connections, such as drinking fountains, water heaters, humidifiers, and air conditioning chillers, be anchored. All water pipes that penetrate construction joints should be provided with adequate flexibility and clearance around the penetration to accommodate the relative movement of the structures. The sprinkler heads on fire suppression systems that penetrate suspended ceilings should be designed with adequate clearance for flexibility to accommodate the movement of the ceiling without damaging or activating the sprinkler head. Plastic sheeting should be stored in rooms that contain electrical equipment so that it can be quickly deployed to protect the equipment from water leaks.

Stem-style suspended light fixtures have frequently failed due to low-cycle fatigue, as shown in Figure 8-3. These units should be provided with secondary safety restraints.

File and storage cabinets should be anchored. Raised computer floors are a nonstructural building component that are discussed in Section 8.5.3, Computer Systems.

8.2 BUILDING SERVICE SYSTEMS

The support systems consist of heating, ventilating, and air conditioning (HVAC) systems and emergency power systems. While both may be essen-

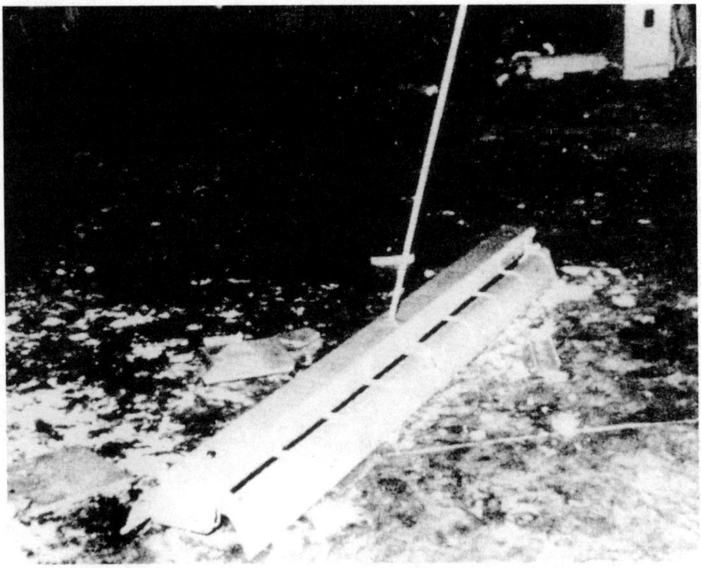

FIGURE 8-3. Stem-Style Light Fixtures Frequently Fail.

tial for continued operation of the control center after an earthquake, the loss of emergency power can be most disruptive.

8.2.1 HVAC Systems

HVAC systems circulate fresh air throughout the facility and provide chilled air to selected areas for personnel comfort and to keep critical equipment, such as computers, cool. Even in cool climates, where air conditioning is normally not used for personnel comfort, it may be needed for computers and communication equipment because of the heat that this equipment generates. HVAC systems consist of chillers, cooling towers, piping, air-handling equipment, heat exchangers, and ducts.

The earthquake performance of HVAC systems has been mixed. Most damage has been associated with poor installation rather than with the equipment itself. The most common problem has been with the roof-mounted cooling towers or heat exchangers. Most commonly, the equipment moves and breaks pipe connections. Flooding resulting from broken water pipes can be very disruptive. Air-handling equipment has also moved (Figure 8-4), but this generally does not affect its operation.

If air conditioning is needed for the continued operation of critical equipment, emergency power should be provided to the system. If evaporative cooling towers are used, the need for makeup water should be evaluated and an emergency supply provided.

FIGURE 8-4. Air Handling Equipment Mounted on Vibration Isolators Moved But Continued To Operate.

8.2.2 Emergency Power Systems

Emergency power is necessary to continue critical functions in the event of a disruption of normal or commercial power. Power system facilities that normally have some type of emergency power include the control center, substations, emergency operations center, and communication facilities.

Two types of emergency power are available to meet system needs. For some systems, no disruption of power is acceptable. For these systems, batteries are used to provide an uninterruptible power supply (UPS). UPS systems may supply power for a few minutes to several hours. In equipment where a short interruption of power is acceptable or when emergency power must be provided for extended periods, an engine-generator system should be provided.

8.2.2.1 Uninterruptible Power Supplies (UPS). UPS systems typically provide emergency power to critical equipment in the event that normal commercial power is disrupted. At substations, station batteries also provide capacity to meet large, short-term power demands. These systems typically consist of storage batteries and racks to support them, battery chargers, and inverters. In some equipment, a distributed UPS system is used in which a battery provides power to an individual equipment rack or to a subsystem in the rack. Some equipment, such as radio repeaters, have very small UPS systems. Figure 8-5 is a schematic diagram of a typical control center UPS system.

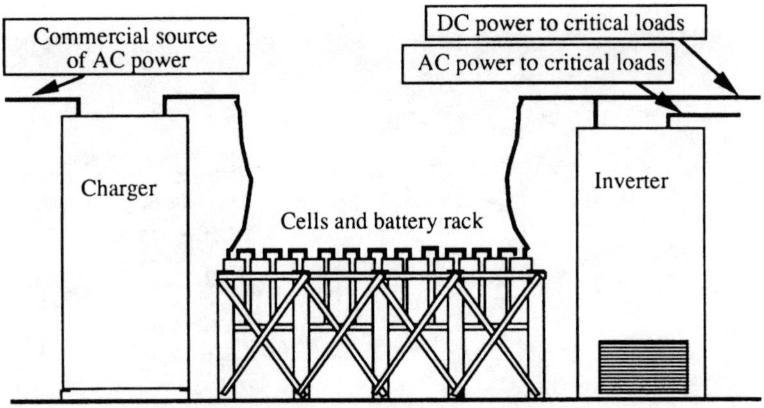

FIGURE 8-5. *Schematic Diagram of UPS System.*

8.2.2.2 ***Battery and Battery Racks.*** The earthquake performance of batteries has been good, provided that they did not fall from their racks. Older cells near the end of their life may have been damaged because of shorting of plates resulting from material having shaken loose within the cell. There has been a report of an internal structural failure in an old cell. The main problem has been damage to cells that have fallen from their racks or been damaged due to impacts while on the rack. Figure 8-6 shows a battery rack, used in some parts of the country, in which the cells are not restrained by the rack. Moderate ground motions can cause the cells to fall from their racks, resulting in cracked cases and acid spills.

Batteries should be restrained by their racks; they should have spacers between cells and between cells and restraints. There is evidence that impacts between adjacent cells and between cells and their restraints can crack cell cases. Heavy cables should be restrained so that cable dynamics do not add loads to connection hardware or binding posts. There should be adequate slack to accommodate rack movement.

Battery racks should be anchored and braced to resist large lateral loads. Bracing a battery rack to a wall can reduce the demands on its structural system and may avoid the need to replace an existing rack. Batteries should be protected from secondary failures, such as damage caused by an unanchored storage cabinet falling on the rack.

8.2.2.3 ***Chargers and Inverters.*** Chargers and inverters are grouped together because they are very similar from an earthquake perspective. Chargers are normally supplied by commercial AC power. This power is converted to the appropriate DC voltage and fed to the batteries, supplying the normal load on the batteries so they remain fully charged. The inverter gets its power from the batteries and converts the DC power to AC power to supply equipment that requires it.

FIGURE 8-6. Battery Rack with No Restraints To Keep the Cells on the Rack.

Chargers and inverters vary in size from breadbox-sized units mounted on a wall to a series of refrigerator-sized units. They should be well anchored because they contain relatively heavy transformers (Figure 8-7). Units have been found that are supported on rails connected to sheet-metal side panels. These units are often supported on short legs or channels with vents in the cabinet floor or at their base to provide good air circulation for cooling. It is important that the load path for the internal transformers be evaluated from the transformers into the foundation pad.

8.2.2.4 ***Special Battery Systems.*** Handheld radios typically have special rechargeable battery packs that are coupled to or inserted into the portable radios. Handheld radios are typically used by maintenance personnel in

FIGURE 8-7. Battery Chargers and Inverters Have Heavy Internal Components That Must Have Adequate Support within the Cabinet.

large facilities, such as power-generating stations. Special battery packs may also be used with flashlights. There have been cases after earthquakes where a power-generating station went black because it lost off-site power and all generating units were off line. After several hours of continuous use, the radio and flashlight batteries ran down, severely hampering the task of getting the units back on line.

The chargers for batteries used in handheld devices usually accept several battery packs in slots. It is important that there be an adequate number of spare battery packs and that the charging unit be connected to a circuit provided with emergency power.

The storage capacity of rechargeable batteries should be periodically checked.

8.3 ENGINE-GENERATOR SYSTEMS

Engine-generator systems are very diverse both in size and configuration. Most systems are designed to start automatically when commercial power is lost, although some systems, especially smaller units, require manual switching for starting. The components and subsystems of a large engine-generator are the engine-generator itself, control console, starting system, day tank, oil cooler, cooling system, main fuel tank, piping system, exhaust system, and transfer switch. A block diagram of the engine-generator shows the functional relationship among the components and subsystems (Figure 8-8). All systems may not have all of these elements. Additional information on the selection, installation, maintenance, and testing of engine-generator systems is contained in endnote 11.

8.3.1 Engine-Generator

Engine-generators vary from small, self-contained units with a capacity of 3 kVA to large diesel units that can generate more than 1,000 kVA. The engine and generator are inherently very rugged. The main vulnerability of the engine-generator is inadequate anchorage. Some units are completely

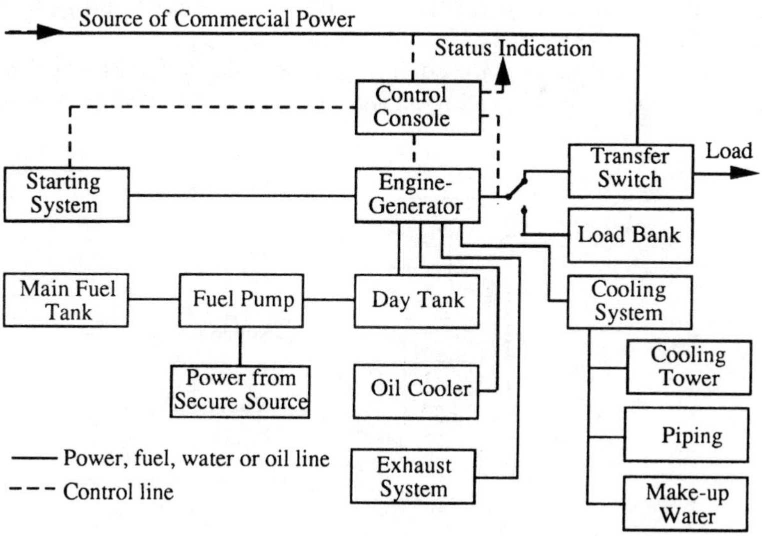

FIGURE 8-8. Functional Components of an Engine-Generator System.

unanchored; more frequently, the engine and generator are mounted on a skid or frame that is supported on vibration isolators to keep vibration of the engine from reaching its support structure. The vibration isolators have frequently failed, allowing the entire system to shift and thus damaging utility connections to the system, such as those carrying water, oil, or fuel. As a result, the engine-generator could not be operated. The vibration isolators should have snubbers to limit the lateral movement of the system (Figure 8-9). Utility line connections to the engine-generator must also be provided with adequate flexibility.

It is recommended that the engine-generator be bolted directly to its foundation. If circumstances require the use of vibration isolators, they should be seismically qualified. In evaluating vibration isolators, cast iron and cast steel can often be found, and the original specifications of the isolators will probably be needed to determine whether snubbers should be added to the unit.

8.3.2 Control Console

Large engine-generators usually have free-standing control consoles. The console controls the sequencing for starting the engine and also has protective systems. The console will have a relay to sense when commercial power is lost, in order to initiate the start-up process. Large engine-generators typically provide power to UPS systems, as well as to many of the other important systems that do not require the continuity of power provided by the UPS. To prevent overloading the engine-generator when the emergency system comes on line, load must be added sequentially. Some loads, such as motors, have a large startup current, which might exceed the generator capacity if all motors were started at the same time.

The protective systems of the engine-generator may include an over-temperature sensor, a low lubricating oil sensor, and associated shut-off relays.

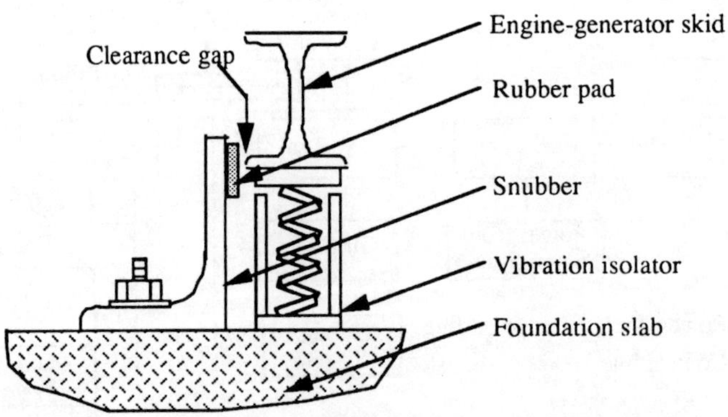

FIGURE 8-9. Snubber Used To Restrain a Vibration-Isolated Engine-Generator.

There have been cases where these relays have been tripped as a result of earthquake vibrations. For this reason vibration-sensitive relays should be avoided. Some facilities may have more than one engine-generator system. In such cases, all engine-generator systems must be electrically synchronized. There have been control system problems when the synchronizing signal was not available.

Some facilities have large UPS systems that can support most functions in the control room. When there is a loss of normal power, system operators may not be aware that the engine-generator did not start—or started and then stopped—until the batteries of the UPS system are almost run down. When this happens, there may be inadequate time to get the engine-generator operating. It is desirable to have an annunciator in the control room indicating the status of facility power and of the engine-generator system.

Free-standing engine-generator control cabinets are typically 20 in. wide but only 6 in. deep. Their small footprint can create large anchor loads due to overturning moments. Braces from the top of the cabinet to a structural member can reduce anchorage loads.

8.3.3 Starting Systems

Two types of starting systems are commonly used for emergency-generators: battery-operated starter motors and compressed air systems. Depending on the size of the engine, one or more batteries may be required to start the engine. Unfortunately, these batteries are often unanchored, so in an earthquake they can move or fall over and be damaged (Figure 8-10). Large engines often use compressed air and air motors to start the engine. These systems require a small electric-motor-operated compressor, an air storage tank, and a piping system. The various components are frequently unanchored, so small-diameter piping can be damaged when equipment moves.

8.3.4 Day Tank

The day tank is a small fuel tank located near the engine-generator that is intended to supply fuel for a short time. (The main fuel tank is located outside of the structure for fire safety reasons.) On some engines, the day tank is supported by the engine. There is usually a fuel pump near or mounted inside the day tank to pump fuel from the main fuel tank to the day tank.

Failures of day tanks have occurred in which their support structure or anchorage failed and fuel oil spilled in the engine-generator room. Because of fear that the oil would be ignited if the engine was restarted, the fuel oil had to be cleaned up before emergency power could be provided.

Environmental Protection Agency (EPA) regulations now require that double-walled fuel tanks be used. A common configuration for the day tanks in these new systems is to place a closed fuel tank inside an open tank,

FIGURE 8-10. A Large Battery Used To Start an Engine-Generator Was Poorly Anchored.

Note: Vibration isolators and snubbers can be seen below the battery.

as shown in Figure 8-11. In these systems, the closed tank may be unanchored or poorly anchored. In this case, the piping is the only physical restraint to seismic loads of the inner tank. Older systems frequently have a separate day tank and fuel pump (Figure 8-12), both often unanchored.

The day tank and associated piping should be anchored. The load path between the internal tank and the foundation should be evaluated. The fuel pump must be connected to a circuit provided with emergency power.

8.3.5 Main Fuel Tank

Most engine-generators have a main fuel tank of 500 to 2,000 gallons. Older tanks were typically buried, but EPA regulations now require that old

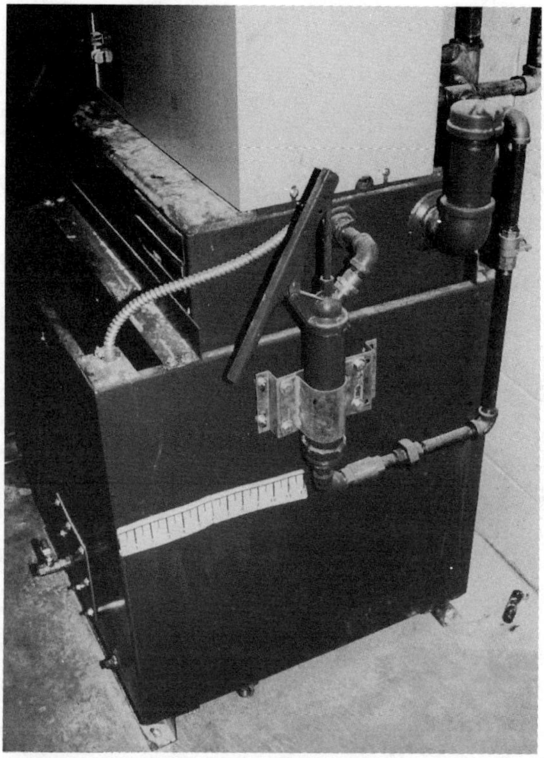

FIGURE 8-11. Day Tank of New Design with Fuel Tank Resting in Open Tank.

Note: While this system is anchored, other units may not have the fuel tank restrained to the open tank. Frequently, the entire system may be unanchored.

tanks be replaced with double-walled tanks, and other requirements favor the use of above-grade tanks, as shown in Figure 8-13. Above-ground tanks are frequently unanchored, and connections can be damaged if they slide (Figure 8-14). One of the problems with large fuel tanks is that the fuel can sit in the tank for a long time before it is either consumed or replenished during routine, periodic testing. Old fuel can clog fuel filters and injectors, preventing the engine from running. While fuel preservatives can extend its life, fuel should not be kept for more than five years, even if it is treated.

Small engine-generators often use bottled-gas as a fuel source. This has the advantages that the fuel will not degrade with time and no day tank is needed. Bottled-gas tanks are frequently inadequately anchored, however, and units have moved in an earthquake and broken fuel lines. Gas leaks in a small enclosure housing the engine can cause an explosion if an ignition source is provided. After a utility experienced an explosion (not related to an earthquake), small ventilation fans and gas detectors were added to gas-

FIGURE 8-12. Older Systems Often Have Separate Fuel Pumps.

Note: The separate fuel pumps are used to move fuel from the main tank to the day tank. They are frequently unanchored, as is the one above.

FIGURE 8-13. Typical Fuel Tank Replacing Older Buried Tanks.

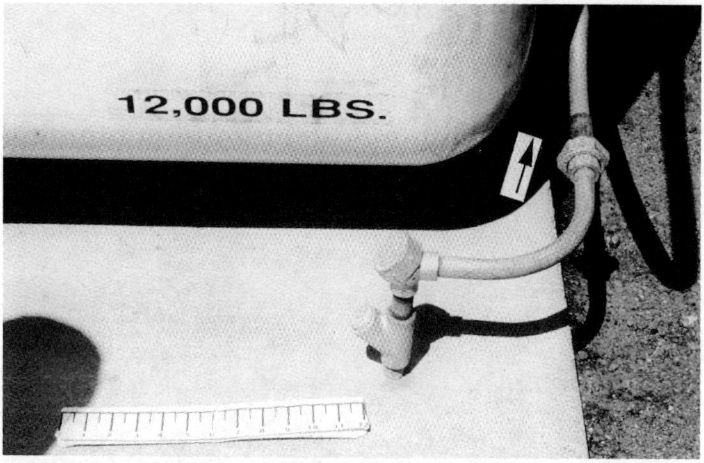

FIGURE 8-14. Above-Ground Tanks Are Frequently Unanchored and Connections Can Be Damaged.

Note: In this instance, anchor tabs are not used and, if the tank slides, it could damage the connection.

fired units. It is important to protect piping from secondary damage. In Figure 8-15, a gas line that supplies the engine-generator could be damaged by an adjacent unanchored storage cabinet.

Emergency power generator systems should not be configured to rely on natural gas supplied from the local gas company. While the earthquake performance of gas system piping has generally been good, distribution lines, service connections, and service lines have been damaged by soil deformations.

8.3.6 Piping Systems

Piping systems are needed to carry the fuel oil from the main tank to the day tank. An unanchored tank can move and cause failure of the fuel line. The main fuel tank is usually located outside of the building containing the engine-generator. Thus, if there is any soil liquefaction or subsidence, the fuel line may be subjected to large strains and fail. There have been cases where the fuel line passed near heavy equipment that moved in the earthquake and fractured the fuel line. There was a case where a long, small-diameter fuel line was inadequately supported, and vibrations caused a low-cycle fatigue failure at a threaded connection.

All pipelines connected to the engine-generator should be evaluated to assure that they have adequate support, that penetrations through construction joints have adequate flexibility, that they are not vulnerable to sec-

FIGURE 8-15. A Small-Diameter Gas Line That Fuels an Engine-Generator Is Vulnerable to Damage from a Nearby Unanchored Storage Cabinet.

ondary failures, and that adequate flexibility is provided to accommodate movement from soil liquefaction.

8.3.7 Oil Cooler

Although uncommon, some large engine-generators have oil coolers. It is important that the lines carrying oil can accommodate the movement of vibration-isolated systems and that the support system for the heat exchanger can withstand earthquake loads.

8.3.8 Cooling System

Almost all engine-generators are liquid cooled. Small and moderate-size engines have radiators mounted on the same skid as the engine-generator. These systems have been the most reliable type of cooling system, because there is no external piping or remote equipment. Larger engines typically have external cooling towers or heat exchangers.

There are two types of cooling towers: closed systems and evaporative-type systems. Closed systems have the advantage that there are no evaporative losses, so no makeup water is needed. Because closed-type cooling systems are more expensive, evaporative-type systems are more frequently used for large engine-generators. Cooling towers are often located on the roof of a building, resulting in increased seismic exposure due to the dynamic response of the building. Engine cooling towers have frequently failed. Some failures were due to inadequate anchorage, for example, when there was a desire to avoid penetrating the waterproof diaphragm covering the roof. In some cases, the anchorage or the structural system making up the cooling tower had corroded over time and failed in the earthquake. There have been cases where piping systems connecting the engine to its remote radiator have failed.

Evaporative cooling towers require makeup water to replace evaporative losses. Because municipal water systems can be disrupted, an emergency water supply should be provided. For small systems, makeup water demands may be moderate, but actual needs should be determined. For example, an insurance company in Des Moines, Iowa, lost their normal water supply due to flooding. The company was unable to use their computers without the air conditioning system to cool them. The cooling of the building could not be isolated from that of the computers. This system required makeup water of 6,000 gallons every eight hours, illustrating that makeup water needs may be substantial.

A third type of cooling system is the single-pass water cooled system. These systems are considered unreliable because they are typically dependent on the municipal water system, which is subject to earthquake disruption, and thus they should be avoided.

Cooling systems in which heat exchangers are mounted on the same support structure as the engine-generator are recommended. If remote heat exchangers are needed, they should have engineered anchorage. It is important that they be provided with proper maintenance to assure their structural integrity. Piping systems between the engine and the heat exchanger should be seismically robust.

8.3.9 Exhaust System

Most engines have a muffler and piping system to remove exhaust fumes from the enclosure protecting the engine. These systems may be poorly supported and fail in an earthquake. There have been cases where the exhaust system was secured to the roof of the engine-generator enclosure and motion of the roof caused the muffler piping connection to fail. While the engine can operate without a muffler, the air intake for the engine is typically in the enclosure and the intake of exhaust fumes from a damaged muffler may cause the engine to be starved for combustible air. Leaking exhaust fumes may also be hazardous to personnel working in the building.

The structural support for the exhaust system should be designed to withstand earthquake loads. If parts of the exhaust system are supported from the roof of the enclosure, the system should be designed to accommodate the relative motion between the engine and exhaust system supports.

8.3.10 Transfer Switch

The transfer switch disconnects the incoming power line from the commercial power source and replaces it with the output of the engine-generator. There have been problems with transfer switches in earthquakes, but they appeared to be associated with issues of age and poor maintenance rather than related to the direct effects of the earthquake. Although the periodic testing procedures used at some facilities do not exercise the transfer switch, it is recommended that the transfer switch be tested at least once a year along with other components, as recommended in the Section 8.3.11, Testing.

8.3.11 Testing

Most organizations periodically test their systems. There have been problems with these systems after an earthquake because the system's controls had not been properly reset after testing or maintenance. There have also been problems with closed air dampers and with closed valves on fuel and cooling system lines.

Periodic testing serves three purposes. First, it establishes that the engine-generator unit(s) can start. Second, it determines that the system can carry its test load. Third, it establishes that, under the test load for several hours, the system does not overheat. Ideally, the system should be tested under full normal load; however, this is often not practical. Many of the systems supported by the engine-generator may not require a UPS system. For example, an inventory control system may not be on a UPS system because the system need not be continuously available. However, the system would be on emergency power following the earthquake because inventory information would be needed. Such systems may be shut down during engine-generator tests. In some cases, a load bank is provided to load the engine-generator during testing to avoid disrupting normal operations. While the load bank may have been appropriately sized at the time the emergency power system was installed, power demand of facilities often increases over time. One of the major causes of failure of engine-generator systems is that they overheat and shut down under actual service loads.

It is recommended that engine-generators be tested during hot weather under normal operating loads at least once a year. The test should run for several hours. It is suggested that, if possible, this test be initiated by cutting normal power to the facility. This procedure will verify the normal start-up process and confirm that all critical subsystems are provided with emergency power.

8.3.12 Operating Procedures and Their Documentation

Some organizations have specialists whose responsibility it is to maintain and test emergency power systems throughout the utility. While this establishes a high degree of expertise and uniformity in maintenance, it is important that additional personnel at the facility be familiar with the operation of the emergency power systems. Should problems arise in the aftermath of an earthquake, local personnel will be in a better position to get the systems into operation. It is recommended that specialists conduct annual training sessions with local personnel in conjunction with the normal testing procedures.

The procedures to be used for starting the emergency power system should be posted near the engine-generator. These instructions should include a check list for all switches and valves associated with the system to assure that they are in the proper position. A schematic diagram should indicate the location of these items and the items should be clearly labeled. A troubleshooting list of potential problems, including symptoms and mitigation measures, should be included. Examples of topics that might be discussed are bad fuel oil, run-down starting batteries, overheating due to overload, overheating due to cooling system damage, and improperly set controls.

8.3.13 Emergency Power Survey

All systems in the facility should be reviewed to determine whether they should be provided with emergency power or connected to a UPS system. As systems evolve, their importance to the operation of the utility may change and warrant their being provided with emergency power. After critical systems have been identified and connected to emergency power, total emergency power demand should be assessed and compared with emergency system capacity.

8.3.14 External Power Hookup

For critical facilities, such as the control center, an external power hookup should be considered. A large telephone company installed external power hookups so that, if there were problems with the engine-generator, a mobile unit could be brought in and quickly connected to the central office. This was done even though an aggressive testing and preventive maintenance program for emergency power systems was in place, because the performance record of emergency power systems has been so poor.

8.3.15 Summary of Critical Issues for Emergency Power Systems

- Engine-generator units should be anchored to their foundations. If vibration isolators are used, they should be seismically qualified, contain no cast iron, and have snubbers or other restraints.

- Engine-generator utility connections should be provided with adequate flexibility.
- The engine-generator control console should be anchored, and top braces should be considered for units with small footprints.
- The status of commercial power and the engine-generator should be provided to the operator through an annunciator.
- Starting system components, such as batteries or a compressed air system, should be anchored.
- The day tank should be anchored.
- The fuel pump, when it is separate from the day tank, should be anchored and provided with emergency power.
- Piping between the main fuel tank, the day tank, and the engine-generator should be secured and provided with adequate flexibility.
- The main fuel tank should be anchored.
- An integral cooling system is recommended. If remote cooling system components are needed, they should be anchored and well maintained, and their piping system anchored and provided with adequate flexibility.
- If evaporative cooling towers are used, the need for makeup water should be evaluated and provisions made for its availability.
- The engine-generator exhaust system support should be engineered, and relative movements of its supports should be taken into account.
- The transfer switch should be properly maintained and its operation tested at least once a year.
- Testing procedures should be established, and testing should be performed on a regular basis. Local personnel should be familiar with the operation of emergency power systems. It is recommended that at least once a year the engine-generator should be tested under normal load conditions with normal power shut off.
- Operating instructions for starting the engine-generator should be posted near the engine-generator.
- An external hookup for mobile emergency power should be provided at critical locations, such as the control center.
- Systems that should be reviewed periodically to determine whether they should be connected to the emergency power system.
- Power demand on the UPS and engine-generator system should be compared to their capacity.

8.4 NETWORK MANAGEMENT

Network management is divided into three activities for this document: energy management, power dispatch, and network configuration.

8.4.1 Energy Management

Energy management, as used here, refers to the contractual relationships between energy producers, consumers, and power transmission companies. While this activity has existed for decades, its scope and character is now rapidly evolving with the deregulation of power systems. The operation of this evolving system and its role relative to system reliability after catastrophic disruptive events, such as earthquakes, is unknown at this time. This activity is dependent on communication among the various players and will utilize the Public Switch Network (PSN) telephone system.

The detailed discussion of the earthquake performance of the PSN is beyond the scope of this document, but is discussed in endnote 12. The earthquake performance of the facilities of the PSN have generally been good. However, immediately after a significant earthquake, call volume dramatically increases and congestion on the network can severely reduce call completion rates. To prevent the saturation of switches (computers) that control network traffic, access into the earthquake-impacted area is restricted by telephone company network control. Under these conditions, communications for the energy management of the power grid may be severely limited.

The traditional method available from most telephone companies was to request telephone lines with essential service status. This is a nontariff service that would be available to some power system telephone lines. This status gives the line priority for getting a dial tone when the system is congested. While this will improve access to the local central office, access to lines outside of the local central office is on an equal basis with all other traffic. As a result, an essential service line status provides only marginal improvement in the availability of the PSN. The Government Emergency Telecommunications Service (GETS) has been designed to address the deficiencies of existing systems. The GETS, in which power utilities would be qualified to enroll, exploits existing facilities of the PSN to provide enhanced routing schemes and priority handling to call completion. Entry into GETS is accomplished by using a special access phone number and a personal identification number (PIN) assigned to each authorized National Security/Emergency Preparedness user. For more information about GETS or how to become a user, contact the National Communication System Home Page on the Internet at http://164.117.147.223 or the GETS PIN Administrator at (703) 607-6118 or via electronic mail at gets@ncr.disa.mil.

It is recommended that essential service status be requested for critical lines within the utility and that appropriate personnel become enrolled in the GETS.

8.4.2 Power Dispatch

Power dispatch is a system that keeps power generation in balance with demand. For most systems, this is a computer control process in which system load and power sources are monitored. Using an appropriate algorithm, the appropriate output from the system's generating units is determined. This information is automatically transmitted to the generating stations and is used to adjust their output. Issues related to the earthquake vulnerability of computers are discussed in Section 8.5, System Control; and communications are discussed in Chapter 9.

8.4.3 Network Configuration

For most power systems, system configuration is supervised by the network operator. Switches may be controlled manually from the local substation or, if motor operated switches are available, remotely from the local substation or the control center. Depending upon the situation at a substation, the substation operator may determine that the substation must be reconfigured, say, to clear a circuit breaker to perform maintenance. These actions would normally be cleared through the control center. The substation operator, if there is one, and the network operator would have the control console change status or enter status if it was not done automatically. Changes in status would be exhibited on the substation and control center status boards. The operator would also have access to a computer monitor from which a broad range of information would be available. Communication systems would also be accessible so that the operator could speak to operators at substations or power-generating stations.

8.5 SYSTEM CONTROL

This section deals with the equipment used for system control. Control consoles, status boards, SCADA systems, and computers are considered.

8.5.1 Control Console and Status Board

The control console usually supports a computer monitor, keyboard, and a communication assets control panel. The console may also support or serve as a bookshelf for reference manuals and an operations log.

The console frequently has a U or V configuration. Power, communication lines, and connection between the monitor and keyboard and the computers usually enter the console through holes in the floor below the console. In some installations, a control board is located adjacent to or behind the control console.

In earthquakes, consoles have moved and monitors, books, and manuals supported on the console have fallen. Because consoles have a low center of gravity, they have not tipped over, but they have slid. The consoles are fre-

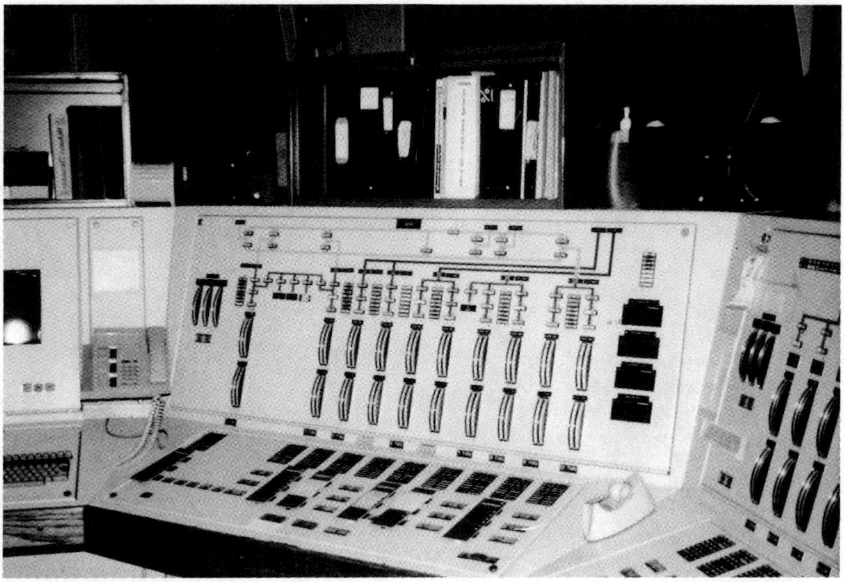

FIGURE 8-16. Control Board Positioned behind a Control Console Operator.

quently poorly anchored. Figure 8-16 shows a control board located behind the control console operator. The books on the top of the control board could fall and activate push buttons on the control panel or damage control levers, switches, or indicators. Figure 8-17 shows a sheet-metal friction clip anchoring a power cabinet, but they have also been used in control councils. The problems with friction clip anchors were discussed in Chapter 5. If the console slides, communication, monitoring, and control lines entering the console through the floor may be cut.

The control console should be securely anchored to the control room floor. Anchoring procedures when raised floors are used are discussed in Section 8.5.3, Computer Systems. Monitors and other equipment should be secured to the control console bench top. This can be done in several ways. While units could be bolted to the surface, the use of benchtop workspace by different operators makes this impractical. The most flexible approach is to use double-sided foam tape to secure the equipment to the surface. This tape, which is available at hardware and stationery supply stores, has a self-sticking adhesive and can be pealed from most surfaces without leaving a residue or damaging the surface; it has a holding power of about 80 pounds per square inch.

Vibration tests of bolted cabinets, such as control consoles and instrumentation racks, have shown that the addition of nuts to bolts using threaded connections can improve their performance.

FIGURE 8-17. Friction Clips Used To Anchor Control Consoles Are Unreliable; if the Console Moves, the Cables Entering through the Floor May Be Cut.

To retain books on surfaces above control consoles, rubberlike shelf-lining materials are now available that provide a high resistance to slipping, leave no residue, and do not stick to the objects on the shelf.

Status boards usually depict the power grid in the region under the control of the control center as a single-line drawing. The status of circuit breakers and disconnect switches is indicated by lights or flags. Status boards may be mounted to structural walls or may be free-standing. Figure 8-18 shows a status board that has been propped up after it fell over in an earthquake.

Status boards should be positively anchored. Self-supporting units with a small footprint should be provided with braces to the building structural system.

8.5.2 SCADA System

Utility supervisory, control, and data acquisition (SCADA) systems monitor information on the status of equipment and values of variables, such as voltage or current, and transfer it to the control center and allow the control center operator to remotely control certain system operations. In addition to the monitoring and control equipment, extensive communication resources are used to transfer information and control signals. Communications are discussed in Chapter 9. The earthquake performance of the terminal parts of

FIGURE 8-18. This Status Board Fell Over in an Earthquake.

SCADA systems has been good. Problems with communications are discussed in Chapter 9.

8.5.3 Computer Systems

The control center may have computer systems for power dispatch, network configuration and status, trouble board damage location processing, system inventory, and power flow and transient analysis. The utility also has computers used for business purposes of billing, accounts payable, and accounts receivable. These computer systems usually require air-conditioned environments and UPS systems to prevent power disruption.

Newer control rooms are often located on raised floors to provide easy access and routing of cables for computers, storage devices, and monitors. The height of raised floors can range from 6 in. to more than 2 ft. The flooring system consists of pedestals, which frequently are held to the subfloor with only a mastic, although shot anchors and expansion anchors are also used. The pedestals are typically positioned on 2-ft centers, and the head of the pedestal supports the corners of four floor panels. Some systems have stringers that connect the pedestals to form a rectangular frame, and, in these systems, the stringers support the edges of the floor panels. Some systems provide structural connections between the stringer and pedestal, while others have pockets for the stringer ends, but no positive connection.

Most equipment is supported on casters or leveling pads (Figure 8-19). This equipment is usually restrained only by the cables that are connected to the equipment and enter through holes cut in the floor panels (Figure 8-20). Heavy equipment is frequently supported on a structural frame that carries loads to the sub-floor.

The most common method of restraining equipment on raised floors is to use the power and signal cables to tether the equipment. These cables can be

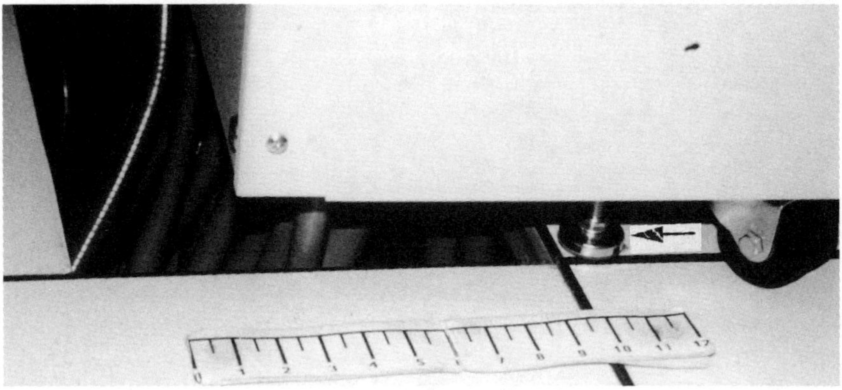

FIGURE 8-19. A Computer Disk Drive Supported on Leveling Screws Adjacent to Hole in Floor for Cable Access.

FIGURE 8-20. Structural Frame Anchored to the Subfloor Supports Heavy Equipment on a Raised Floor.

secured to the floor slab. In some cases, steel cable is attached to the equipment frame and anchored to the floor slab to reduce the loading on power and signal cables and their connectors. In both of these systems, very little vertical restraint is provided, so tall equipment may tend to tip over. Equipment is typically free to slide several inches, enabling it to hit walls, columns, or adjacent equipment. Casters or leveling pads can also fall into cable access holes.

8.5.3.1 Earthquake Performance of Computer Systems. Most earthquake damage has been the result of unanchored equipment moving, tipping, or sliding into adjacent equipment, walls, or building columns. In systems that do not have stringers connecting to pedestals, there have been partial collapses of the raised floor system. If equipment is anchored to the raised floor, then substantial lateral loads can be transferred to the raised floor. Panels can pop up and the members can buckle, if there are no stringers or the stringers do not have structural connections. Equipment supports can fall into cable access holes and tip the equipment over. Unanchored equipment can strike walls or columns. In a Japanese computer facility using similar installation practices, computer tape decks fell over (Figure 8-21).

There have been water leaks, and water has collected below the raised floor; as a result, cable connections have become wet and been damaged.

8.5.3.2 Recommended Practices for Computer Systems. Pedestals that support the floor panels should have positive anchors. Mastic does not have adequate holding capacity, but high-strength bonding material would be acceptable. Shot-installed nails should be avoided because they pull out as a result of the prying action of the pedestals subjected to overturning moments. Positive connections should be used between the stringers and pedestals.

Equipment on raised floors should be restrained. The use of power or signal cables to tether equipment is not recommended since it can damage cables and does not prevent the equipment from tipping over or moving some distance laterally. If this approach is used, a separate tether cable should be attached to the equipment and secured to the subfloor.

A toggle-link system has been used to secure equipment to raised floors. In this system, a heavy bar is connected across the center of the equipment frame base. An expansion anchor eye bolt is installed in the floor slab beneath the equipment. A hole is cut in the floor panel, and a rod and turnbuckle are connected between the eye bolt in the floor slab and the bar in the equipment to form a toggle link (Figure 8-22). With this system, the equipment is allowed to slide on the raised floor, but the toggle link holds the equipment down and limits its lateral movement. The hole in the floor for this toggle link must be large enough to accommodate lateral movement of the equipment. With this method, some movement of the equipment is

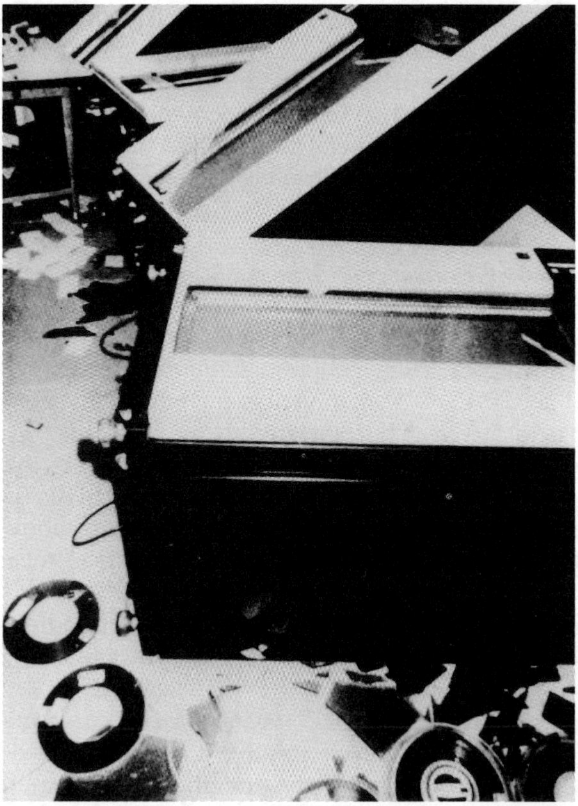

FIGURE 8-21. Computer Tape Decks That Fell Over in a Japanese Facility.

Note: Signal and power cables restrained lateral movement at the base of the computer tape drive. Lacking restraints at the top and vertical restraint at the base, the tape drive fell over.

allowed, but there are no impacts at the limits of the motion. Relatively little lateral load is applied to the raised floor. As the equipment moves, the load in the toggle link increases, preventing the equipment from tipping over.

Another approach is to attach a steel cable at each corner of the equipment, pass these through small holes in the floor, and anchor them to the floor slab (Figure 8-23). This system provides lateral and vertical restraint. It has performed well, but the small holes in the raised floor for the steel cables will transfer lateral loads to the raised floor if the equipment moves.

Heavy equipment should be supported on a frame that is anchored to the floor slab (Figure 8-20).

Each of the three methods of restraining equipment on raised floors has advantages and disadvantages. The use of tethers has been tested on shake tables; however, the peak-to-peak motion during the test was limited to 3 in.

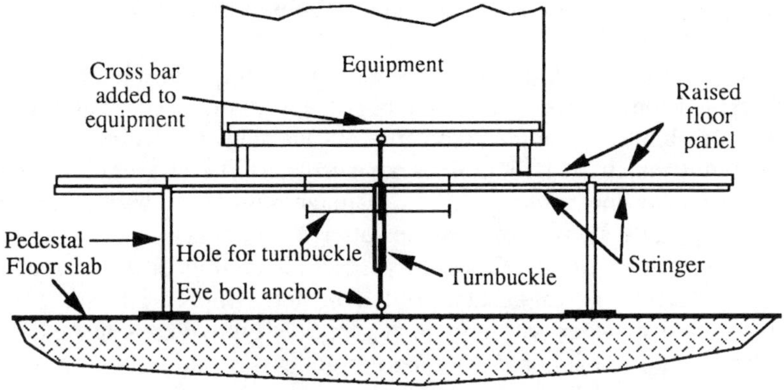

FIGURE 8-22. Toggle-Link System for Restraining Equipment on Raised Floors.

FIGURE 8-23. Steel Cables Have Been Attached to Each Corner, Passed through Holes in the Raised Floor, and Anchored to the Subfloor.

by the table limitation. Much larger motions can be anticipated in earthquakes, so there are questions about the applicability of the tests to more realistic base motions. The use of a frame anchored to the floor slab is generally considered the most effective, but it is also the most costly.

8.5.4 Damage Restoration

8.5.4.1 Damage Assessment. In general, after an earthquake, substation personnel will assess damage, and engineers from the central office will review each substation to document damage. Pictures of damage before cleanup starts can be useful for postearthquake evaluation of the equipment failure modes and factors that may have contributed to the failures. A detailed procedure for conducting a post-earthquake evaluation of power systems is given in endnote 13.

Damage to the distribution lines is assessed by inspection by field crews and from customer calls. The trouble board, a term not used by all utilities, consists of a bank of operators who answer calls made by customers reporting system problems. Many utilities have centralized this operation for the entire system at one or two locations. Damage reports can then be entered into a computer system, and, from the pattern of outages, the location of the problem can be identified. Once the problem is identified, the centralized system will send a message that is printed out at the appropriate service center, where a dispatcher will send an appropriate type of crew to repair the problem. (The trouble board operation in general would not be physically located at the control center, but is described in this section because of the centralized character of its operation.)

8.5.4.2 Damage Locator. Earthquake damage to substations is determined by postearthquake inspections. Damage assessment to the distribution system would be done in the same way as is done for storm damage. Customer calls to the trouble board are used to locate the damage, which is then followed up by dispatching a crew to identify and repair the damage. Congestion on the Public Switch Network telephone system and the flood of calls to the trouble board can slow the receipt of damage reports at the utility. This can frustrate customers. Experience indicates that customers are sympathetic to the problem of restoring service after a damaging earthquake; however, customer frustration can turn to anger if reliable information is not provided as to when service is to be restored. Some utilities have developed sophisticated damage locator systems. The availability and use of these systems can aid in providing timely restoration predictions to customers.

8.5.4.3 Repair Dispatch and Service Centers. Distribution system repair notices are given to the appropriate service center based on customer calls. The service center contains a supply of spare parts, parking for service vehicles, design drawings for the service area, communication links, and a base station for dispatching repair crews.

There is no record of severe damage to service center structures. There has been damage to spare porcelain that was not properly stored. Small parts stored on shelves and in storage cabinets have been tossed on the floor. While this did not physically damage most items, the time taken to find the desired parts slightly delayed the restoration process. At one service center, an unanchored water heater located on the second floor broke a water pipe connection. The water leaked onto communication equipment on the first floor, and all utility communications were disrupted for several hours. At another service center, communications were disrupted for some time after the earthquake, apparently because disruption in the utility microwave station occurred after the system was checked for damage by service personnel.

The earthquake vulnerability of the service center buildings should be evaluated, and vulnerable structures should be scheduled for upgrading or replacement within the overall earthquake mitigation program. The recommendations for storing porcelain members are contained in Chapter 5.

8.5.4.4 Damage Records. In the aftermath of a damaging earthquake, normal operating procedures are often bypassed in an effort to quickly restore service. Verbal work orders replace their computer or paper counterpart. Spare parts are often removed from storage without requisitions. While the restoration of service takes top priority after personnel safety, resources should also be devoted to documenting restoration costs. For most publicly owned utilities, the Federal Emergency Management Agency (FEMA) can reimburse the utility if there is a presidential declaration of a disaster. For investor-owned utilities, losses will have to be documented for insurance claims.

8.5.4.5 Special Governmental Services. Under certain circumstances, federal agencies can provide assistance for investor-owned utilities. For example, in the Northridge earthquake, FEMA made arrangements for military air transports to carry heavy substation equipment needed for restoration. These services can be made available at no cost to the utility if public health and safety are served by the action.

Once the President declares a disaster, the Federal Response Plan is activated. Under this plan, the Department of Energy can make available emergency fuel supplies if normal supplies are disrupted.

ENDNOTES

1. Federal Emergency Management Agency. 1988. *Rapid Visual Screening of Buildings for Potential Seismic Hazards: A Handbook.* FEMA 154.
2. Federal Emergency Management Agency. 1988. *Rapid Visual Screening of Buildings for Potential Seismic Hazards: Supporting Documentation.* FEMA 155.
3. Federal Emergency Management Agency. 1992. *NEHRP Handbook for the Seismic Evaluation of Existing Buildings.* FEMA 178.
4. Federal Emergency Management Agency. 1989. *Establishing Programs and Priorities for the Seismic Rehabilitation of Buildings: Supporting Report.* FEMA 173.
5. Federal Emergency Management Agency. 1989. *Establishing Programs and Priorities for the Seismic Rehabilitation of Buildings: A Handbook.* FEMA 174.
6. Federal Emergency Management Agency. 1992. *NEHRP Handbook of Techniques for the Seismic Rehabilitation of Existing Buildings.* FEMA 172.
7. Federal Emergency Management Agency. 1992. *A Benefit-Cost Model for the Seismic Rehabilitation of Buildings, Vol. 1: A User's Manual.* FEMA 227.

8. Federal Emergency Management Agency. 1992. *A Benefit-Cost Model for the Seismic Rehabilitation of Buildings, Vol. 2: Supporting Documentation*. FEMA 228.
9. Federal Emergency Management Agency. 1994. *Typical Cost for Seismic Rehabilitation of Existing Buildings, Vol. 1: Summary, Second Edition*. FEMA 156.
10. Federal Emergency Management Agency. 1995. *Typical Cost for Seismic Rehabilitation of Existing Buildings, Vol. 2: Supporting Documentation, Second Edition.* FEMA 157.
11. A.J. Schiff, Ed. Forthcoming. *Guide to Reliable Emergency Power.* American Society of Civil Engineers, Technical Council on Lifeline Earthquake Engineering.
12. A. Tang and A.J. Schiff, Eds. 1996. *Methods of Achieving Improved Seismic Performance of Communications Systems.* American Society of Civil Engineers, Technical Council on Lifeline Earthquake Engineering, Monograph 10.
13. A.J. Schiff, Ed. 1997. *Guide to Post-Earthquake Investigation of Lifelines.* American Society of Civil Engineers, Technical Council on Lifeline Earthquake Engineering, Monograph 11.

Chapter 9

COMMUNICATION SYSTEMS

Recommendations: Retrofitting and New Construction

Communication equipment racks, including the main distribution frame, should be provided with stiff base anchorage details and braced at the top to cable trays or walls (9.2.1).

Cable tray connections should be made to structural elements with bolted connections; J-bolts should not be used (9.2.2).

Cable drops between cable trays and equipment should be provided with adequate slack (9.2.2).

PBXs should be anchored to their supports using the manufacturer's recommendations (9.2.3).

Procedures for anchoring equipment on raised floors are discussed in Section 8.5.3.2.

Circuit boards in communication equipment should have positive restraints to keep the cards in their card cages (9.2.3).

Modems should have positive anchorage to their supports (9.2.4).

The installation of waveguides on microwave equipment should be made in order to limit loads on waveguide connections to the equipment rack (9.2.5).

Radio equipment at base stations and at repeaters should be anchored (9.3).

Emergency power should be provided to critical communication systems; seismic issues related to emergency power are discussed in Chapter 8 (8.2.2.2 through 8.2.2.4; 8.3.1 through 8.3.10).

HVAC systems may be needed to maintain the operation of communication systems; seismic issues related to HVAC systems are discussed in Chapter 8 (8.1.2, 8.2.2, 8.3.13).

Recommendations: New Construction

Cable tray sections should be joined with bolted connections (9.2.2).

Utilities typically have several communication systems. In this document, these systems have been divided into three groups: communication links, communication equipment located at facilities, and radio-based maintenance dispatch systems.

9.1 COMMUNICATION LINKS

Microwave and optical fiber systems are typically used to form the links of a utility's communication system. Some utilities also use leased lines from telephone companies.

Before the introduction of optical fiber systems, microwave systems formed the backbone of utility communications. These systems link the control center with generating stations and major substations. In addition to the terminal facilities, which are discussed in the next section, repeater stations are required. These microwave stations form line-of-sight links between facilities. Repeater stations consist of receivers, transmitters, antennae, a source of commercial power, and emergency power. They are often located at high points along the route so that the distance between stations is maximized and the number of stations reduced. Optical fiber has been introduced more recently. Fibers are typically carried on transmission towers within a shield wire or are supported on the tower as a separate optical-fiber cable. In some cases, the optical fiber is buried along tower rights-of-way or other routes.

The earthquake performance of microwave towers has generally been good. Wind loads control the structural design, and the seismic loads are relatively small. Microwave dishes are designed to be adjustable, and there are cases in which adjustments slipped during earthquakes and the resulting misalignment caused a loss of signal strength or loss of signal. There are reports that communications have been temporarily disrupted during earthquakes; this has been attributed to misalignment due to the dynamic response of the tower. In the 1992 Philippine earthquake, a steel lattice microwave tower about 15 m tall and similar in design to those used by the power industry failed. In the 1995 Kobe, Japan, earthquake, a large telecommunication tower mounted on top of a central office building was severely damaged and eventually removed. This tower was much larger than those used by utilities. A few members of a microwave tower at a substation in Kobe buckled, but operations were not affected. There are no documented failures of optical fiber communication links in earthquakes: however, failures in telephone company fiber cables were reported in the Kobe earthquake.

Special concerns about repeater stations will be discussed in Section 9.3, Radio-Based Maintenance Dispatch.

Carrier systems using power transmission lines have been used for communications, but these systems are typically limited to system protection. These systems were discussed in Chapter 5, Substations.

9.2 COMMUNICATION EQUIPMENT LOCATED AT FACILITIES

Figure 9-1 is a block diagram depicting communication equipment that might be found at a control center or substation. Radio equipment used for mobile units is not shown in this figure. Not all facilities have all equipment as shown, and the size and form of the equipment can vary between facilities. Large utilities often have a utility-owned telephone system with equipment similar to that found in a telephone company central office. The equipment can be grouped into power, support, telephone, microwave equipment, and communication links. Telephone and microwave equipment are discussed in this section.

9.2.1 Telephone Equipment

The main distribution frame can be a large rack (Figure 9-2) or a small board mounted directly on a wall for a small facility. Communication racks

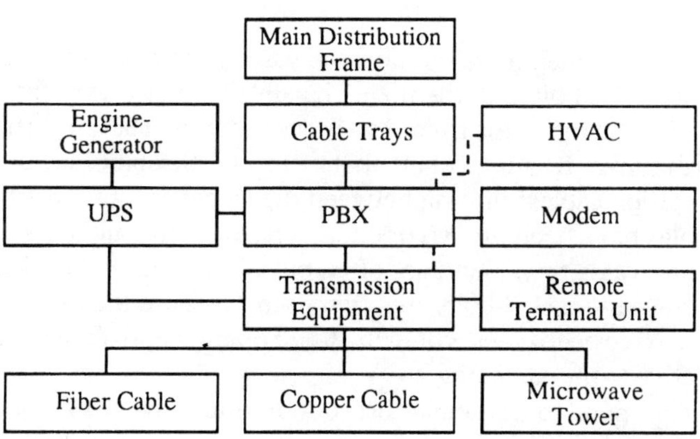

FIGURE 9-1. Communication Equipment Found at Control Centers or Substations.

FIGURE 9-2. Main Distribution Frame at a Power System Facility.

Note: Large rack deflections have been observed due to the flexible base detail.

typically have a small footprint and the base anchorage is formed by two aluminum angles bolted to the floor. This detail yields a very flexible structure in which the top of the frame can move more than 30 cm (1 ft) in a moderate earthquake. It is important that adequate flexibility be provided to communication cables running between the frame and the overhead cable trays. Cables have been broken or cut as a result of the movement between the rack and cable tray. Racks are often bolted to overhead cable trays that serve to brace the rack. J-bolts are often used to secure the rack to the cable tray. There have been cases where the J-bolt opened and allowed the rack to move relative to the cable tray.

It is recommended that main distribution frames be braced at the top to limit movement and overturning moments on the anchor bolts. The load

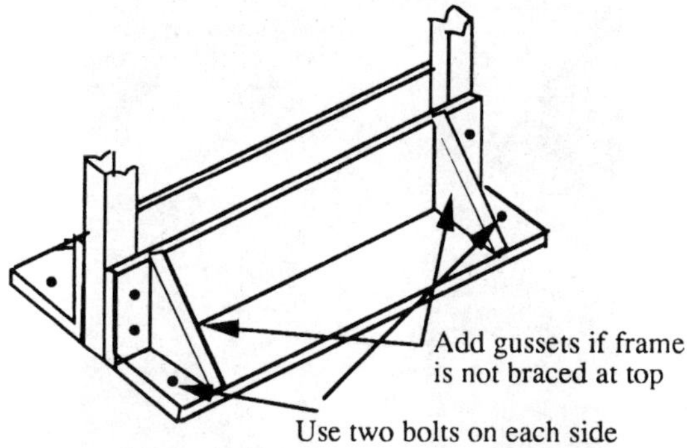

FIGURE 9-3. Communication Rack Anchorage Detail.

path of the bracing system should be engineered. Instead of J- and U-bolts, bolted-through connections should be used to secure equipment. Four anchor bolts, two on each side of the rack, should be used. Racks without top bracing should have gussets placed on the base angles to stiffen the anchorage (Figure 9-3). Such gussets can be subjected to large loads and have been observed to fail in shake-table tests if they are not properly installed.

9.2.2 Cable Trays

Cable trays used in the communication industry typically are ladder-type trays. Sections are joined by friction clips (Figure 9-4) that provide very little load capacity in tension when used to brace equipment racks. Cable trays are often anchored to walls with the use of J-bolts, which can open in an earthquake (Figure 9-5).

It is recommended that power-system-type cable trays be used for communication equipment. These trays have bolted connections between cable tray sections. If communication-type cable trays are used, sections should be joined with bolted connections. J-bolts should not be used as structural connections. The performance of communication-type cable trays used as braces can be improved if they span from wall-to-wall. Cable drops from cable trays to equipment should be provided with ample slack to accommodate relative motion between the tray and the equipment.

9.2.3 Private Branch Exchange (PBX)

A modern PBX is a specialized computer that connects calls made between telephones within the system and to the Public Switch Network. Figure 9-6 shows an unanchored PBX at a power facility. In an earthquake, it could tip over and damage cable connections or internal components.

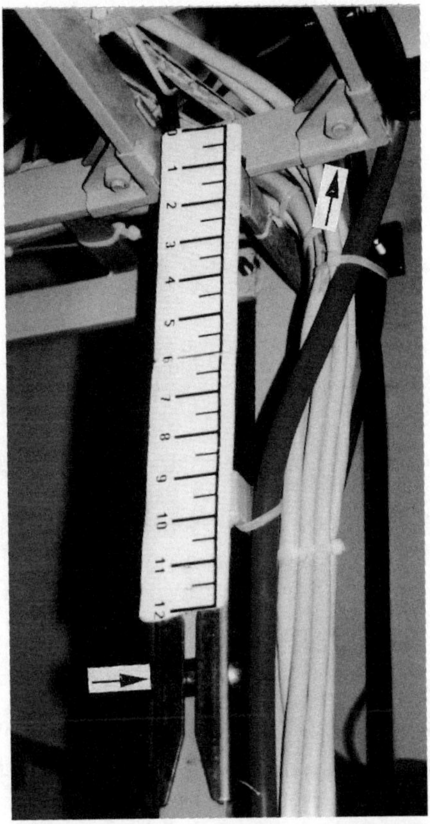

FIGURE 9-4. Friction Clips Are Typically Used To Connect Cable Tray Sections.

It is recommended that the PBX be securely anchored to the floor following manufacturer's recommendations. Issues related to anchoring equipment to raised floors are discussed in Chapter 8. Modern PBX equipment is designed for base anchorage, so bracing at the top is not needed. The PBX should also be provided with an uninterruptible power supply. Following a power loss, PBXs must be rebooted and their software reloaded. For large systems, this can be a lengthy process.

PBXs and other communication equipment typically are constructed with a card case and motherboard for a back plane. Circuit boards are often plugged into the motherboard with only the friction of the connector for restraint (Figure 9-7). There are many examples of the circuit boards vibrating out of their card cages in earthquakes. Most modern communication equipment that has been seismically qualified for the communication industry has clips to restrain the circuit boards. Older equipment has clips that are used to remove the circuit board but do not restrain it.

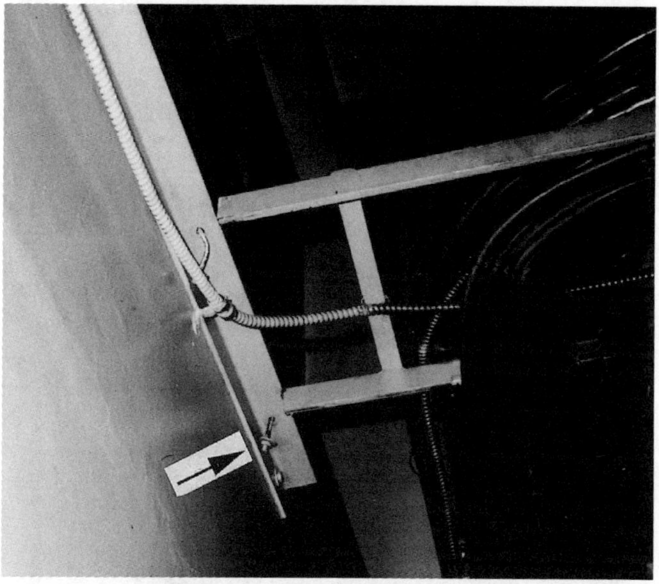

FIGURE 9-5. *J-Bolts Are Often Used To Secure Cable Trays to Structural Walls.*

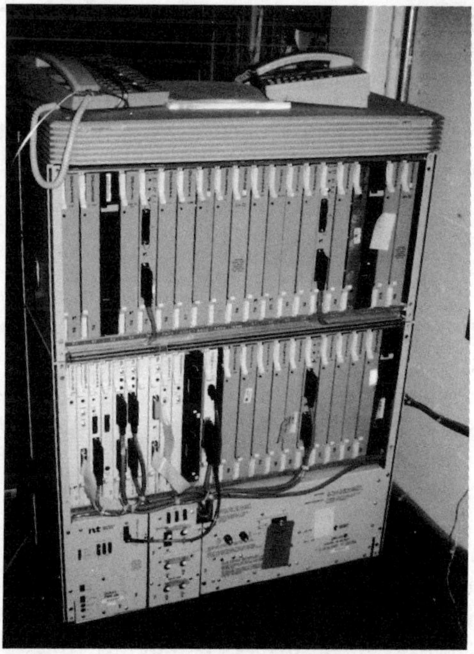

FIGURE 9-6. *Telephone Switch Used in a PBX System at a Power-Generating Facility; the Switch Is Not Anchored.*

FIGURE 9-7. Circuit Boards Are Not Positively Restrained in Their Card Cage.

Circuit boards should be restrained in their card cages. Figure 9-8 shows a simple method for retrofitting equipment to restrain circuit boards.

9.2.4 Modems

Modems are devices that provide an interface between various types of equipment, such as monitoring equipment, and telephone communication channels. The devices are typically cigarbox size and are often added to an existing installation. As a result, they are seldom properly anchored (Figure 9-9). These units are lightweight, so that they are unlikely to be damaged or to damage other items themselves if they fall. However, the relatively fragile ribbon cables that are often used to make communication connections are vulnerable.

Modems should be anchored. Because of the modems' small mass, double-sided tape can provide adequate restraint. (A double-sided adhesive foam tape is available as a consumer product; its the adhesive is self-sticking and can be pealed from most surfaces without leaving a residue or damaging the surface.) Tapes with adequate capacities should be used; some foam tapes have rated capacities of 80 pounds per square inch.

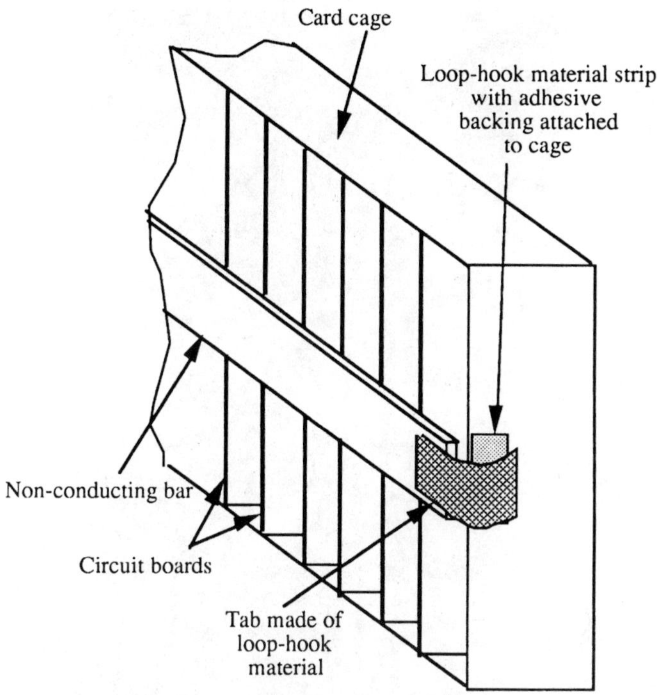

FIGURE 9-8. Method To Keep Circuit Boards in Their Card Cages.

9.2.5 Microwave Transmission Equipment

Microwave equipment is typically mounted in communication equipment type racks. Special care is needed in the installation of waveguides, because of the flexibility of these racks. Figure 9-10 show a typical rack at a substation. The battery at the base of the rack provides power to the UPS that supports equipment in the rack. The rack is anchored to the cable tray above it.

Figure 9-11 shows a waveguide on a microwave rack at a utility control center. The rack is not connected to the cable tray, and the waveguide passing through the cable tray has only about 0.5 in. clearance with the tray. Waveguides are assembled with small machine screws. In some cases, nylon machine screws are used (Figure 9-12). As a result, the load capacity of the waveguide connections in the rack is limited. Figure 9-13 shows a waveguide that penetrates a T-bar suspended ceiling. In an earthquake, the motion of the ceiling could cause waveguide connections to fail. After leaving the rack, flexible waveguides are typically used and they are much less susceptible to damage.

Care must be exercised in the installation of a waveguide to limit loads at waveguide–equipment connections. The flexibility of the equipment rack must be considered.

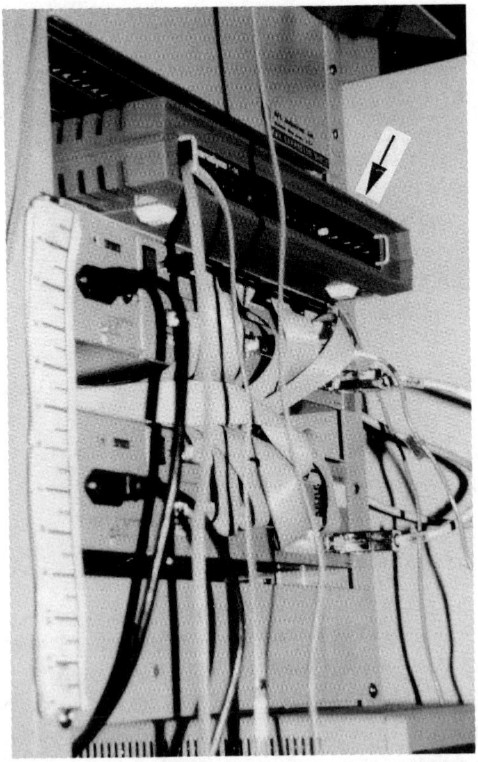

FIGURE 9-9. Modems Are Seldom Properly Anchored.

9.3 RADIO-BASED MAINTENANCE DISPATCH

Utilities make extensive use of radio systems to communicate with maintenance personnel within large facilities, such as power plants, and with personnel in service vehicles. For large utilities, several regional systems may be used. Each will have a base station, usually operated from one of the service centers, and repeaters located on ridges and hilltops to get good coverage over the service area.

There is no record of damage to base station equipment. At one installation, the signal cable between the antenna support structure and an adjacent structure containing the transmitter broke because of a lack of flexibility in the connection. Communication equipment should be anchored.

Repeaters are often located on ridges and hilltops, and they often can be accessed only by steep, unimproved roads. Access can be impeded by earthquake-induced landslides. Also, commercial power to some sites is often vulnerable because of very steep access routes.

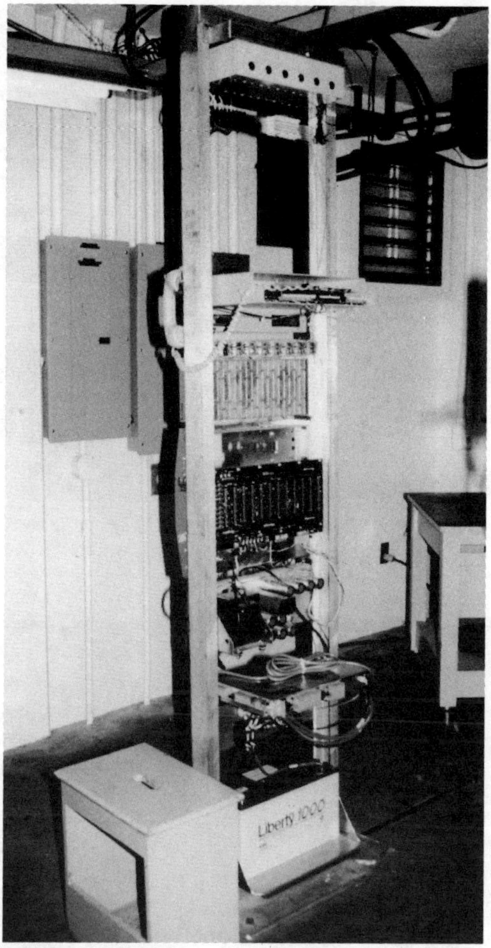

FIGURE 9-10. Microwave Equipment Is Typically Mounted in Communication Equipment Type Racks.

Earthquake motions on ridges and hilltops can be much more severe than in surrounding areas; thus, very conservative anchorage should be used at repeater sites. In the Loma Prieta earthquake, dozens of nonutility repeaters went off line. Many units were unanchored and fell to the floor. Commercial power at many repeater sites was disrupted, and many sites had only battery backup power and that lasted for only a few hours. The performance of utility repeaters is not known.

A large utility generally has one or more frequencies allocated to it for its exclusive use. There have been communication problems with incompatible

FIGURE 9-11. Relative Movement between the Rack and the Cable Tray Could Damage Waveguide Connections.

FIGURE 9-12. Small Screws, in This Case Made of Nylon, Require That Earthquake Loads on Waveguides Be Limited.

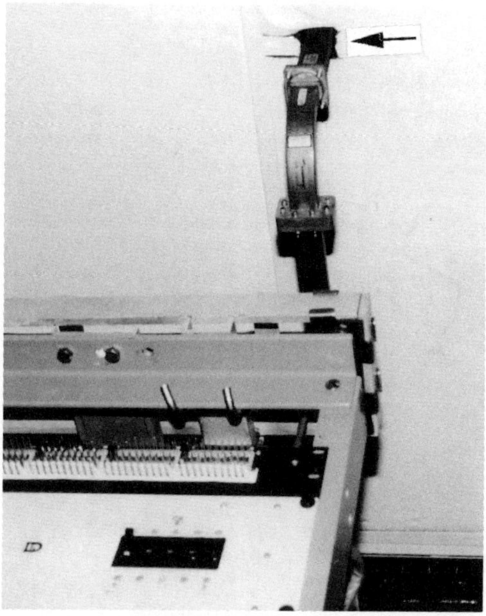

FIGURE 9-13. Motion of the Suspended Ceiling Could Induce Loads on the Waveguide and Damage Connections.

frequencies in service vehicles brought in from other districts or from other utilities under mutual aid. Under these conditions, radio systems have become overloaded after earthquakes. Supervised network control methods are sometimes helpful to improve communications.

9.4 EMERGENCY POWER

Most communication equipment must remain functional after an earthquake, even if power is disrupted. Communication equipment should be provided with emergency power in the form of UPS or other backup power systems. Seismic issues related to emergency power are discussed in Chapter 8.

9.5 HVAC SYSTEMS

Many communication systems require a temperature-controlled environment for proper operation. Such systems need HVAC systems for this purpose. The HVAC systems may also require emergency power for continued operation if there are power disruptions. Seismic issues related to HVAC systems are discussed in Chapter 8.

Chapter 10

ANCILLARY FACILITIES AND FUNCTIONS

Recommendations

Structures that house the service centers, inventory control system, spare parts, and the emergency operations center should be evaluated for earthquake vulnerability and postearthquake serviceability; their renovation or replacement should be included in the mitigation program (10.1, 10.2, 10.3, 10.4).

The installation of equipment needed for postearthquake operations should be done using good seismic practices (10.1, 10.2).

Equipment that requires power and is needed for postearthquake operations should be provided with emergency power (10.1, 10.2, 10.4).

An update of the inventory of critical supplies not included in the inventory control system, such as spare high-voltage transformer bushings, should be made (10.2).

Spare parts should be stored so that they can survive an earthquake. Parts cabinets should be anchored and braced. Large items stored on the floor or in storage rack should be secured to shelves to prevent their damage. (10.3)

Critical parts, such as high-voltage bushings, should be completely bolted to anchored support structures (10.3).

Expedited methods for tracking supplies and spare parts after an earthquake are needed (10.3).

The utility's emergency response plan should include an Emergency Operations Center (EOC). The seismic vulnerability of the EOC building should be assessed. (10.4)

An external emergency power hook-up for a mobile generator should be provided to the EOC and energy control center (10.4).

EOC communications using the Public Switched Network should request essential service lines, and the utility should take advantage of the Government Emergency Telecommunications Service (GETS) system (10.4).

Periodic exercises of the emergency response plan should include the EOC to evaluate its organization and effectiveness (10.4).

Arrangements should be made for competent postearthquake inspection of utility buildings, particularly those needed for continued operation (10.5).

Remote alternative sources should be identified for specialized equipment normally rented locally (10.6).

In this Manual, the term *ancillary facilities and functions* is used to describe maintenance service centers, inventory control systems, spare parts storage, emergency operation centers, engineering offices, and specialized equipment. Specialized equipment includes large cranes, cherry pickers, and low-boy heavy-equipment transporters used by utilities to relocate equipment after an earthquake.

The documentation of the earthquake performance of ancillary facilities is relatively poor. Damage to or impaired operation of these facilities and functions usually does not directly impact power system service. However, after a severe earthquake in which extensive damage overwhelms system redundancy, the operation of ancillary facilities, such as service centers or engineering department offices, and other support functions would almost certainly extend the recovery period.

10.1 MAINTENANCE OR SERVICE CENTERS

Service centers play a vital part in the restoration of electrical service after a damaging earthquake through their role in the dispatch of service crews and allocation of spare parts and repair equipment. Often, service centers serve as storage depots. The buildings that house service centers can have wide variations in their seismic capacity, even within a single utility.

There have been earthquakes in which the operations of one or more service centers have been severely disrupted by direct effects of the earthquake on service center structures, service center communication equipment, and spare parts. In one earthquake, the loss of commercial power and the lack of emergency power severely disrupted the operation of a service center. With the loss of power, there were no lights, the computers that provided access to the inventory database were down, the base station radio for dispatching service vehicles was down (a mobile unit in a maintenance vehicle was used, but it did not have the range of the base station and the use of that vehicle to make repairs was lost), and air conditioning was down, so telephone and microwave equipment became overheated. For unexplained reasons, all telephones, including the utility-owned system at the service center, were out of service for several hours after the earthquake. In this case, damage had to be reported using the overloaded and impaired radio system. In addition, loss of commercial power to repeaters impaired the dispatching system.

In one service center, a small, unanchored water heater above the first floor, which was used to heat water for sanitary facilities, tipped over and broke its pipe connections. As a result, water leaked onto communication equipment on the first floor, knocking out all communication with the service center.

Experience shows that service center equipment is generally inherently seismically rugged, and it performs well, if it is properly installed.

It is recommended that the seismic vulnerability and postearthquake serviceability of service center structures and their equipment be evaluated and that their renovation or replacement included in the mitigation program. Communication and other support systems, such as HVAC equipment and equipment used to link to inventory databases, that are needed for continued operation should be provided with emergency power. This equipment should be installed so that it survives earthquakes undamaged. Seismic issues related to emergency power and communications are discussed in Chapters 8 and 9, respectively.

10.2 INVENTORY CONTROL SYSTEMS

Most utilities have computer-based inventory control systems that have evolved over the last 20 to 30 years. These systems may be vital for determining the availability and location of replacement parts needed for postearthquake restoration of service and recovery. These systems may also be needed for ordering replacement equipment from manufacturers.

Most of these systems contain information on equipment and supplies acquired since the installation of the inventory system. Spare parts, such as spare bushings for transformers, that predate the inventory system may not be included. With the recent downsizing of utility staff, the individuals who know the locations of spare parts not in the inventory system may no longer be available.

The seismic vulnerability and postearthquake serviceability of inventory control systems and of the terminals needed to access the databases should be evaluated, and their renovation should be included in the mitigation program. These systems should also be provided with emergency power. Seismic issues related to computer systems and emergency power are discussed in Chapter 8. The mitigation program should also include a review of critical supplies that may not be included in the inventory database.

10.3 SPARE PARTS STORAGE

Spare parts play an important role in system restoration after a damaging earthquake. Spare parts may be stored at warehouses, at service centers, or at the facilities where they are to be used (that is, substations or power-generating stations).

Warehouses are often metal-clad structures, which have performed well in earthquakes, or tilt-up structures, which have a mixed seismic record. Within warehouses, very large stored items may rest on the floor, while other materials are often stored on industrial shelving (Figure 10-1). Typically, items stored on shelves are unsecured and vulnerable (Figure 10-2).

The storage of spare parts at service centers is similar to that at warehouses or at substations and the same precautions should be taken in their storage.

Substations typically have a limited number of general purpose supplies, of spare parts needed at the substation, and of specialized parts for equip-

FIGURE 10-1. Large Industrial Racks Are Used in Warehouses To Store Equipment.

FIGURE 10-2. Large Items on Shelves Are Vulnerable If They Are Not Secured.

ment at the substation. Spare parts are often stored with no consideration given to earthquakes. For example, small parts are often stored in general purpose cabinets (Figure 10-3). Spare parts cabinets are often left with doors open, and shelves typically do not have retainer lips. The cabinets are often unanchored. High-voltage bushings may be stored in special racks (Figure 10-4). Even when bushings are stored in special racks, they may not be bolted to the racks and the racks may be unanchored. Some bushings are stored in an ad hoc manner (Figure 10-5). Large bushings may weigh several tons, and the exposed rope restraining them can deteriorate in the sun and provide little restraint. Original shipping cases are often used to store bushings and post insulators (Figure 10-6). These are adequate and perform well; however, when exposed to the elements, they quickly deteriorate and provide little protection for their contents. The storage of porcelain members at some substations is well planned and executed (Figure 10-7).

The loss of spare bushings can be particularly disruptive because they are one of the more seismically vulnerable items and their loss can cause prolonged delay in repair of transformers. Smaller parts stored on shelves have been dumped to the floor. Although this typically caused little damage, it delayed the availability of the inventory for use in restoration.

In an effort to expedite the restoration of power, the work-order system is often bypassed. This has led to a loss of inventory control, so that the number and location of spares are not known. There are examples of needed spare parts being delivered to the wrong site, where they sat unused while they were needed at another location. There have also been cases where work crews removed items from inventory and hoarded them in anticipation of their needs.

FIGURE 10-3. A Small Parts Cabinet at a Substation Was Unanchored and the Doors Were Left Open.

It is recommended that the seismic vulnerability and serviceability of storage facilities be evaluated and their renovation or replacement be included in the mitigation plan. Storage cabinets should be anchored and braced near the top of a structural wall or member. Base anchorage alone is not recommended because the sheet metal used to fabricate the cabinets may not withstand anchorage loads on a heavily loaded cabinet. Storage shelves should be provided with lips to keep items on the shelves. Heavy items stored on the floor or on industrial storage racks should be secured to prevent damage. Fortunately, fragile spare parts, such as bushings, have relatively little inventory turnover, so that securing them is not overly burdensome. Bushings held in racks should be secured with all flange bolts and the racks should be anchored.

FIGURE 10-4. Bushings Can Be Stored in Racks at Substations, but the Racks Are Often Poorly Anchored.

An expedited method for tracking spare parts after an earthquake is needed.

10.4 EMERGENCY OPERATIONS CENTERS

An effective emergency operations center (EOC) can significantly improve the emergency response and recovery of a utility after a destructive earthquake. Federal regulations require that power systems have emergency response plans, and these plans usually call for the creation and operation of an EOC. The EOC is often located near the utility's main operations center. Because of the importance of EOCs to utility operations in a

FIGURE 10-5. Large Bushings Are Secured to a Structure with Rope That Could Deteriorate in the Sun.

postearthquake environment, many utilities also have an alternative EOC at a remote location, in the event that the primary facility is damaged or becomes inaccessible.

The role of the EOC is to gather key information on emergency response, to assess damage, to coordinate and prioritize restoration activities within the utility, and to provide information about utility restoration status to the public and various agencies. The EOC assesses the need for resources and attempts to make arrangements to mitigate shortages. In general, the EOC has administrative authority over the utility, but most operational decisions are delegated to facility managers and maintenance operating personnel.

In general, a room or group of rooms is designated to serve as the EOC. While these rooms may be solely for the use of the EOC, more often the areas used by the EOC serve other functions when the EOC is not activated. The EOC is primarily dependent on communications and power resources. In some facilities, water, sewage, and gas lifelines also play a role. The seismic vulnerability of the building that houses the EOC is also of concern.

FIGURE 10-6. Porcelain Spare Parts Are Often Stored in Their Shipping Crates, Which Quickly Deteriorate if Exposed to the Weather.

FIGURE 10-7. Storage Area in Which All Porcelain Members Are Well Anchored.

Most organizations have established criteria in their emergency response plans to describe circumstances under which the EOC is to be activated. These plans usually have procedures established to summon EOC staff to the center via pager or telephone. In severe events where communications may be disrupted, the need to report is usually obvious; this should be noted in the disaster response plan.

10.4.1 Organization of an EOC

Each utility generally has a unique organization for its EOC to meet the special needs of the utility. Many are modeled after the fire service Incident Command System. Within California, the state has mandated the use of a standard terminology and organizational structure to improve priority setting, interagency coordination, and the efficient flow of resources and information. This system is referred to as the Standardized Emergency Management System (SEMS). The SEMS has five *functions*, as indicated in Figure 10-8. It also has five *organizational levels* as indicated in Figure 10-9.

Pacific Bell has been one of the pioneers in utility emergency response planning. The description of a utility EOC given below has drawn heavily

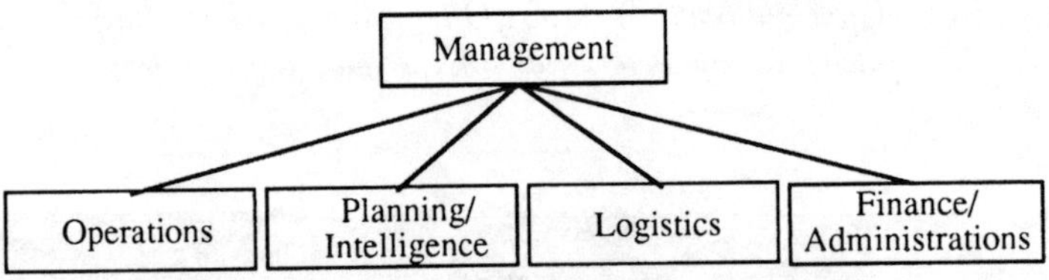

FIGURE 10-8. Five Functions of the SEMS.

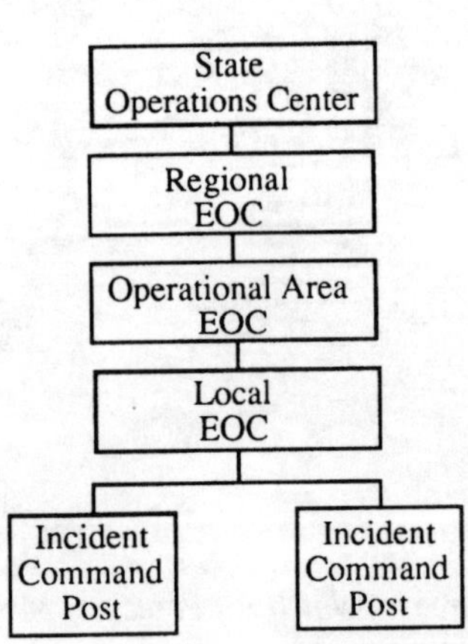

FIGURE 10-9. Five Organizational Levels of the SEMS.

from material provided by Pacific Bell. This general approach has been adapted by at least one large power utility. Clearly, many of the sections and the specific nomenclature in Pacific Bell's organization chart are unique to a telephone company. However, most functions provided for will be needed in EOC organization plans of power systems. It should be noted that while a utility EOC may have administrative authority over regional or division operations, traditionally operating units are given a large degree of autonomy and authority over their operations.

The company emergency response plan designates the roles and responsibilities of various groups, including the Company Emergency Policy Committee and Emergency Operating Center Teams. The Company Emergency Policy Committee provides policy direction in support of the EOC teams and service restoration teams' efforts. The EOC Teams have preplans that support assigned functions that include assessing damage, prioritizing and coordinating restoration, allocation of resources, and communication of information and status to utility groups, governments, employees, media, and other stakeholders. EOC Team members should be prepared to work at a remote EOC site. The organizational chart for the EOC is given in Figure 10-10.

The physical layout of the EOC is organized around four tables that correspond to four of the columns in the organizational chart (Figure 10-11). The functional names of white boards are distributed around the walls of the main EOC room. The functions associated with each table and with other functions identified in the organizational chart are discussed below. Note that items marked with an asterisk identify the terms used in California's SEMS.

10.4.1.1 The Four Table/Team Concept.

- Command and control within the EOC
- Damage report assessment
- Plans and coordinates restoration activities
- Prioritizes and allocates resources
- Keeps employees and customers informed

10.4.1.2 Liaison Table (Operations).*

- Receives damage assessment information, updates, and field requests for resources and assistance
- Functional point of contact in the EOC
- Posts information received on Information Category Status Boards
- Team Leader forwards requests for assistance to the Team Leader at the Coordination Table
- Team Leader represents Liaison Table in all Strategy Team meetings
- Team Leader is responsible for ensuring current preplans at table

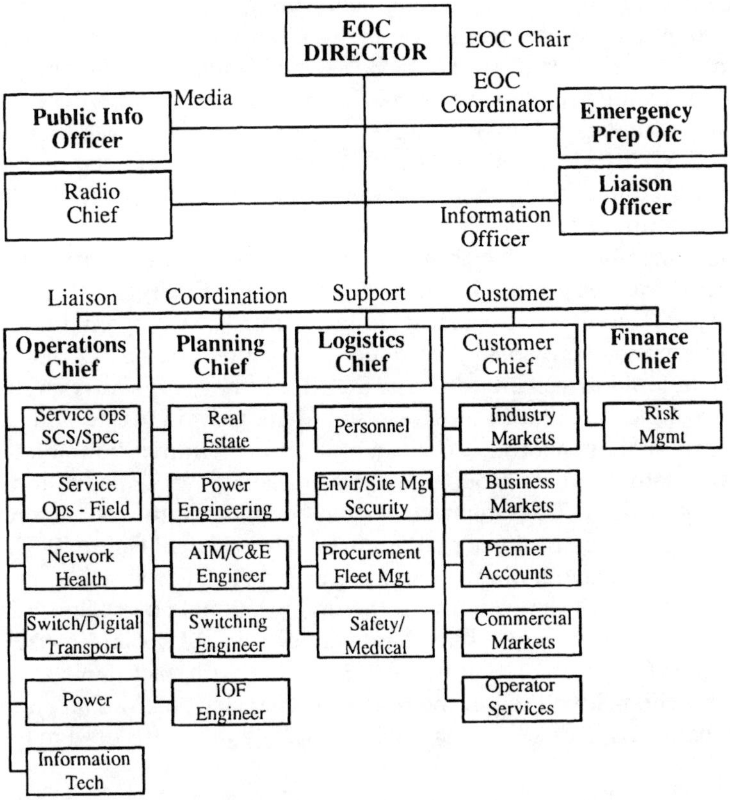

FIGURE 10-10. Schematic Diagram of a Telephone Company's EOC.

Note: Bold print reflects the EOC titles as set forth under California's State Emergency Management System.

10.4.1.3 Coordination (Project) Table (Planning).*

- Members act as project managers/subject matter experts (SME) for specific requests received from the Liaison Team
- Posts EOC Project Board for table responsibilities
- Project Managers coordinate requests with the Support Table Team Leader or members as appropriate
- Team Leader represents Coordination Table in all Strategy Team meetings
- Team Leader is responsible for ensuring current preplans at table

10.4.1.4 Support Table (Logistics).*

- Assists Coordination Teams with internal/external resources and support

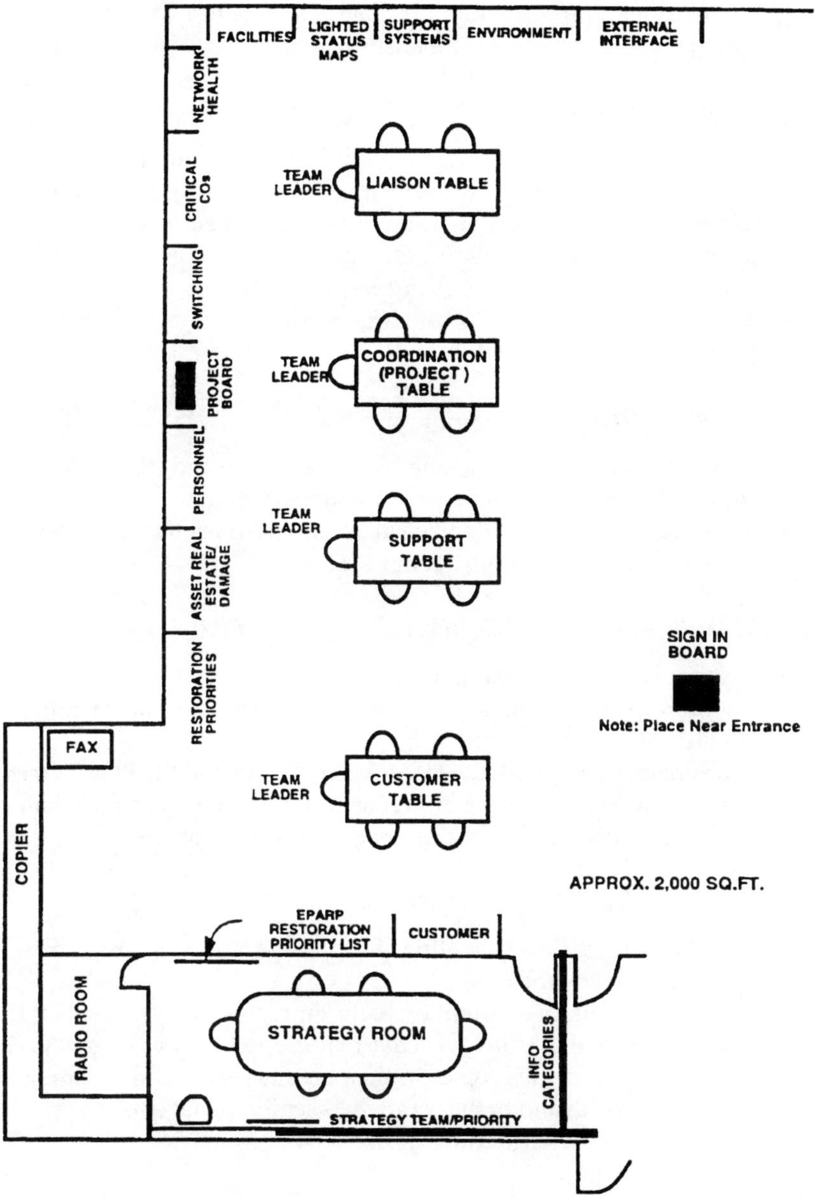

FIGURE 10-11. Generic Physical Layout for a Telephone Company's EOC.

- Posts information on status boards as appropriate, such as Personnel and Environmental boards
- Team Leader represents Support Table in all Strategy Team Meetings
- Team Leader is responsible for ensuring current preplans at table

10.4.1.5 Customer Table.

- Team members interact directly with customers and customer-interfacing organizations to support and prioritize specific restoration efforts and to advise customers on the status of their service
- Team Leader interfaces with EOC members as necessary for understanding of customers impact and restoration activity
- Team Leader represents Customer Table in all Strategy Team meetings
- Team Leader is responsible for ensuring current preplans at table

10.4.1.6 Finance Team.

- Team is responsible for tracking costs of material and labor associated with response to a disaster and restoration of service
- Team tracks cost of asset damage in a disaster for insurance purposes
- Team may participate from a remote location

10.4.1.7 Media Manager (Public Information Officer, PIO).

- Is responsible for providing the broadcast and print media with information regarding company's situation, the customer impact, and restoration information
- Coordinates closely with the Information Officer and the EOC Director
- Provides media interviews or prepares and accompanies the Chair or Vice President when interviews are deemed appropriate

10.4.1.8 Information Officer (Liaison Officer).

- Is responsible for consolidating the damage and customer requirements received by the EOC
- Prepares a report for distribution to the emergency control center (The ECC is an office that is always active to respond to any emergency. The ECC, in turn, provides copies to the necessary state and federal government agencies and to the company's senior management.)
- Coordinates closely with the EOC leadership team

10.4.1.9 Strategy Room.

- Typical representation includes:
  Chairperson – (EOC Director*)

Information Officer (Liaison Officer*)
EOC Coordinator (Emergency Preparedness Officer*)
Liaison Team Leader (Operations Chief*)
Coordinator Team Leader (Planning Chief*)
Support Team Leader (Logistics Chief*)
Customer Team Leader
External Affairs/Media (Public Relations Manager*)
Vice President

- Summarize current status from damage/customer reports
- Prioritizes restoration based on:
  Information Categories
  Restoration Priority List
- Reports status to ECC, state and federal government, officers, and media
- Team Leaders update respective Teams after each strategy meeting

10.4.1.10 Radio Room.

- Qualified radio operators establish wireless communications in the event of disruption to the land lines
- Radio nets are monitored, and pertinent information collected is provided to the EOC Team

10.4.1.11 Communications Using the Public Switch Network (Regular Telephone). The Public Switch Network is the primary means for the EOC to communicate outside of the utility. The EOC is often fitted with extra lines and jacks, and handsets are moved into the room when the EOC is activated. Commercial lines in EOCs can be given essential service status in some telephone systems, which will give them priority access to the dial tone when the telephone system is congested. Implementation practices of the essential service feature are a function of the telephone company and the equipment they use. Once a phone is connected to its central office, the call is not distinguished from other calls in getting access to external trunks. Organizations that are part of the Federal Emergency Response Plan will generally have access to the Government Emergency Telecommunications Service (GETS). GETS is a White House-directed and federally funded service that supports the National Security and Emergency Preparedness community. GETS provides enhanced routing schemes and priority treatment features; it has been established by the major long distance carriers and is being implemented by some local telephone companies. With this service, priority routing is provided for the completion of calls. It is important that the EOC has access to GETS. Users are provided with a special telephone card with a personal identification number that gives them access to the system.

10.4.2 Past Earthquake Performance of EOCs and Similar Facilities

Three types of problems have been frequently encountered:

1. earthquake performance of the structures that house EOCs has been mixed;
2. communications into and out of EOCs have often been impaired; and
3. emergency power has failed when commercial power has been disrupted.

Structures that house the EOC have been severely damaged and have had to be evacuated. Some structures have been red-tagged by building inspectors and have had to be evacuated.

Key to the operation of the EOC is communication with utility facilities and with outside organizations. There have been cases where telephones were difficult to use because of congestion on the Public Switched Network. In some cases, essential service lines had dial tone delays as long as three minutes. In many facilities, the telephone lines were not placed on essential service status, so there were long delays in obtaining dial tone. There have also been situations where the EOC was not provided with an adequate number of lines to operate effectively.

The loss of commercial power is fairly common in severe earthquakes. The EOC is typically provided with emergency power. However, emergency power has frequently been unreliable. Many factors have contributed to the failure of emergency power and these are discussed in Chapter 8. Because of the large number of factors that can contribute to emergency power disruptions, some organizations have installed an external power hook-up to facilitate the use of external mobile generators.

10.4.3 Interaction of the Utility EOC and Government EOCs

After a damaging earthquake, there will usually be a federal disaster declared at the request of the affected state(s). This triggers the Federal Disaster Response Plan, and governmental EOCs will be activated. Many situations arise that require the utility EOC to communicate with governmental EOCs and, possibly, to respond to requests made by them. Likewise, governmental EOCs may be able to respond to requests made by utilities. For example, after the 1989 Loma Prieta earthquake, military aircraft, which have very large load capacities, were used to deliver large power equipment to the damaged area to expedite the restoration process.

The state of California may have a unique situation. In 1952, the governor chartered the California Utilities Emergency Association (CUEA). The CUEA was created by a Joint Powers Agreement to represent California utilities on utility emergency-related issues. The CUEA is a voluntary membership association of utilities, including electric, gas, pipeline, telecommunications, water, wastewater, and pipeline utilities. It provides a structure for communication and coordination among government agencies and public- and investor-owned utilities throughout the state. Funding is provided primarily by dues from member utilities. The CUEA operates and manages the Utilities Branch of the Governor's Office of Emergency Services. The executive director of the CUEA also serves as the chief of the Utilities Branch.

The CUEA has an EOC (known as the Utility Operations Center, or UOC) within the state EOC in Sacramento. When activated, it is staffed by the Utilities Branch Chief, and member representatives are provided by CUEA member utilities.

In addition to operating the UOC when it is activated, the CUEA has ongoing planning and educational activities to enhance the emergency response of all utilities.

10.4.4 Recommendations

An EOC should be part of the utility's emergency response plan. The EOC should be provided with emergency power. The installation of an external emergency power hook-up for a mobile generator should be provided. EOC communication using the Public Switch Network should request essential service lines, and the utility should take advantage of the GETS system. Periodic exercises of the emergency response plan should include activating the EOC to evaluate its organization and effectiveness.

10.5 ENGINEERING OFFICES

The utility's engineering office plays a vital role in the long-term restoration of a utility after a damaging earthquake. If the damage overwhelms the redundancy designed into the system, the engineering staff will play an important role in the restoration of service. Engineering drawings and equipment specifications may be needed for the restoration of facilities and should be available.

The seismic vulnerability of the structure housing the engineering offices is important because of the safety of utility personnel and accessibility after the earthquake. There have been cases where buildings have had to be evacuated because of structural damage or the building was red-tagged after the earthquake. That is, many communities have established programs in which buildings are inspected by the building department after a damaging earthquake. Buildings can be tagged as safe or in need of additional inspection or unsafe. Building departments often mobilize inspectors, primarily structural engineers, from outside the building department to help in the massive inspection process. The need to quickly evaluate buildings and a tendency to be conservative can result in sound buildings being tagged as unsafe. There have been cases of utilities being denied access to a rented building by the building owner who is concerned about liability.

It is recommended that seismic vulnerability and postearthquake serviceability of engineering department buildings be evaluated and that renovation or replacement be included in the mitigation program. In addition, the emergency response plan should include a plan for competent postearthquake inspection of the utility's structures through the use of utility engineers or of local engineering firms who are on retainer. FEMA has published documents to aid in postearthquake evaluation of buildings.[1,2]

10.6 SPECIALIZED EQUIPMENT

A utility often needs specialized equipment to recover after a damaging earthquake. Some specialized equipment, such as large cranes and low-boys to move heavy equipment and helicopters for surveying lines, is usually rented. Some equipment, such as cherry pickers, high potential testers to check for internal damage to transformers, and equipment to condition transformer oil, may have to be rented to supplement equipment owned by the utility. While retainers may be paid to assure the availability of such equipment, these contracts may be preempted by government emergency services organizations. There may be no economical solution to this loss of service, but response plans should take into account that these services may not be available locally. Some of this equipment is mobile, so that it can be brought in from distances of more than 1,000 km within 24 hours.

ENDNOTES

1. Federal Emergency Management Agency. 1988. *Rapid Visual Screening of Buildings for Potential Seismic Hazards: A Handbook.* FEMA 154.
2. Federal Emergency Management Agency. 1988. *Rapid Visual Screening of Buildings for Potential Seismic Hazards: Supporting Documentation.* FEMA 155.

Appendix A

MODIFIED MERCALLI INTENSITY SCALE

The following scale is derived from C.F. Richter's *Elementary Seismology.*

I Not felt.

II Felt by persons at rest, on upper floors, or in favorable places.

III Felt indoors. Hanging objects swing. Vibration like passing of light trucks. Duration estimated. May not be recognized as an earthquake.

IV Hanging objects swing. Vibration like passing of heavy trucks or sensation of a jolt like a ball striking the walls. Standing motor cars rock. Windows, dishes, doors rattle. Glasses clink. Crockery clashes. In the upper range of wooden walls and frames creak.

V Felt outdoors; direction estimated. Sleepers wakened. Liquids disturbed, some spilled. Small unstable objects displaced or upset. Doors swing, close and open. Shutters, pictures move. Pendulum clocks stop, start, change rate.

VI Felt by all. Many frightened and run outdoors. Persons walk unsteadily. Windows, dishes, glassware broken; knickknacks, books, etc. fall off shelves. Pictures fall off walls. Furniture moved or overturned. Weak plaster and masonry D cracked. Small bells ring in churches and schools. Trees, bushes visibly shaken or heard to rustle.

VII Difficult to stand. Noticed by drivers of motor cars. Hanging objects quiver. Furniture broken. Damage to masonry D, including cracks. Weak chimneys broken at roof line. Fall of plaster, loose bricks, stones, tiles, cornices (also unbraced parapets and architectural ornaments). Some cracks in masonry C. Waves on ponds; water turbid with mud. Small slides and caving in along sand or gravel banks. Large bells ring. Concrete irrigation ditches damaged.

VIII Steering of motor car affected. Damage to masonry C; partial collapse. Some damage to masonry B; none to masonry A. Fall of stucco and some masonry walls. Twisting, fall of chimneys, factory stacks, monuments, towers, elevated tanks. Frame houses moved on foundations if not bolted down; loose panel walls thrown out. Decayed piling broken off. Branches broken from trees. Changes in flow or temperature of springs and wells. Cracks in wet ground and of steep slopes.

IX General panic. Masonry D destroyed; masonry B seriously damaged. (General damage to foundations.) Frame structures, if not bolted, shifted off foundations. Frames racked. Serious damage to reservoirs. Underground pipes broken. Conspicuous cracks in ground. In alluviated areas, sand and mud ejected, earthquake fountains, sand craters.

X Most masonry and frame structures destroyed with their foundations. Some well-built wooden structures and bridges destroyed. Serious damage to dams, dikes, embankments. Large landslides. Water thrown on banks in canals, rivers, lakes, etc. Sand and mud shift horizontally on beaches and flat land. Rails bent slightly.

XI Rails bent greatly. Underground pipelines completely out of service.

XII Damage nearly total. Large rock masses displaced. Lines of sight and level distorted. Objects thrown into air.

QUALITY OF MASONRY

To avoid ambiguity, the quality of masonry, brick, or other material is specified by the following system. (This has no connection with the conventional classes of A, B, and C construction.)

Masonry A. Good workmanship, mortar, and design; reinforced, especially laterally, and bound together by using steel, concrete, etc.; designed to resist lateral forces.

Masonry B. Good workmanship and mortar; reinforced, but not designed to resist lateral forces.

Masonry C. Ordinary workmanship and mortar; no extreme weaknesses, like failing to tie in at corners, but neither reinforced nor designed to resist horizontal forces.

Masonry D. Weak materials, such as adobe; poor mortar; low standards of workmanship; weak horizontally.

Appendix B

INVESTIGATION OF SOIL-STRUCTURE INTERACTION OF ELECTRICAL EQUIPMENT

This appendix has been abstracted from parts of an unpublished report prepared by Dames & Moore for the Los Angeles Department of Water and Power (DWP) in 1996[1]. This abstract will focus on the modeling and evaluation of a single-phase transformer at the Rinaldi Receiving Station but will comment on a transmission tower.

The analysis utilized the FLAC, Version 3.3, computer code to evaluate the seismic performance of the soil-structure system of the transformer, foundation, and surrounding soil. FLAC is a two-dimensional, explicit, finite difference code to model geotechnical problems. The model required several simplifications of the soil-structure system to economize the analysis and reflect physical limitations about soil data obtained from field exploration and laboratory tests. The transformer model only included the transformer case and high- and low-voltage bushings. Details of the conservator and radiators were not represented. The moments experienced at the base of the bushings served as the performance criteria. The soil and structure model is depicted in Figure B-1.

The soil nonlinear effective-stress static and dynamic deformation analysis was performed using elastoplastic constitutive laws based on Mohr–Coulomb failure criterion. The elastoplastic law was coupled with an incremental shaking-induced, pore-water pressure generation scheme based on work done by Professor Seed in the 1970s. Material properties for soil were bases on previous investigations and a literature survey on data from soils with similar characteristics.

The slope of the surface within the site generally did not exceed 1%. Known subsurface explorations within the site have generally been limited to shallow pits and Cone Penetration Tests. Using the available data, three representative soil profiles were established. In the first analysis profile, the upper 10 ft of the site consisted of loose to medium-density silty sands and

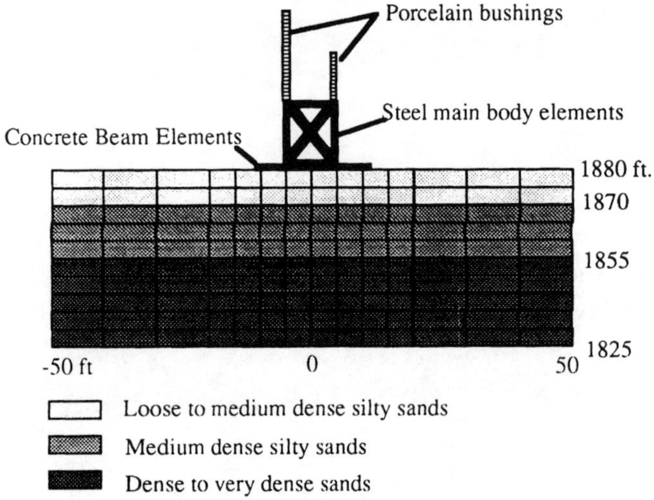

FIGURE B-1. Soil Model of the First of Three Typical Soil Profiles Extended to 55 ft below Grade

sandy silts. The next layer consisted of a 15-ft thick layer of medium-dense silty sands. Below the medium-dense soil, dense to very dense sands were encountered to the maximum depth explored, about 50 ft. Groundwater was not encountered at the site. A DWP report indicates a groundwater table at 160 ft below grade at the site. The soil model shown in Figure B-1 is for the first of three typical soil profiles extended to 55 ft below grade, where the soil was considered stiff enough to act as bedrock.

The transformer body was 15-ft high and supported a 17- and a 10-ft-high porcelain bushing. The transformer was supported on a 2.5-ft thick concrete slab.

A free-field strong motion record at the site was used in the analysis.

Several analyses were performed, each proceeding until failure was observed or ground motion subsided. The porcelain bushing failed 2.5 seconds into the record. Tests using the two alternative soil profiles did not significantly alter the results. To investigate if shallow groundwater affects the soil-structure interaction, an additional analysis assuming a 20-ft depth to groundwater was conducted. There was no significant change in the soil-structure interaction. This study substantiated that soil-structure interaction, rather than liquefaction, was the cause of transformer rocking at the site.

Two approaches to remediation were evaluated. In the first, a lean concrete slurry grout was injected to a depth of 15 ft below and adjacent to the transformer foundation, as shown in Figure B-2. The grouting did not change the outcome, and the porcelain still failed. In the second approach,

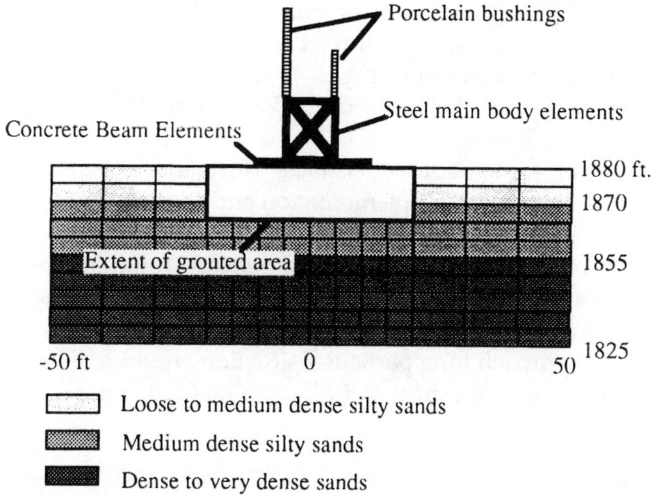

FIGURE B-2. Soil Model of the First of Three Typical Soil Profiles Extended to 55 ft below Grade with Grouted Area Shown.

2-ft-diameter drilled piers extended to depth of 15 ft. Piers were installed on 5-ft centers to support the foundation slab. This approach reduced moments on the bushings to just below their ultimate load. The small margin of safety may indicate that this approach is inadequate.

Another evaluation used "infinite strength" porcelain to determine the peak load experienced by the bushings. This evaluation showed the peak bushing load was 120% of the ultimate strength. There were no foundation failures in any of the evaluations.

The report also evaluated three other structures at a different substation that had a different soil profile in which there was liquefaction at the site. The remediation approaches were the same: use of a grout to improve the soil around the existing foundations and use of piers to support the foundation. In all cases where the piers did not penetrate into the firm lower strata, the remediation improved the performance but did not prevent failure. In the case of a transmission tower, where the piers penetrated the lower competent soil, the performance was acceptable. That is, if the tower foundation pier was only in liquefied soil, it would be pulled from the soil.

Based on results of the study, the following conclusions relative to transformers can be made.

- Soil-structure interaction can increase seismic loads on transformer bushings.
- Grouting, to the extent used in the study, does not significantly improve earthquake performance.

- The use of piers that do not extend into competent soil can improve earthquake performance, but for the situation studied, the bushings were still on the verge of failing.
- The use of piers that extend to competent soil can significantly improve performance.
- The use of piers to improve seismic performance is a costly mitigation procedure in terms of implementation and an evaluation to substantiate that it is an effective approach.
- The use of more robust bushings in general would be the most cost-effective approach; however, there may be cases where such bushings are not available.
- The best approach for a particular situation would require an evaluation to determine the most cost-effective method of mitigation.

ENDNOTE

1. Dames & Moore. 1996. "Final Report: Investigation of Soil-Structure Interaction, Electrical Equipment at Sylmar Converter Stations and Rinaldi Receiving Station, Sylmar, California." Prepared for the Los Angeles Department of Water and Power, Job No. 00138-092-015.

Appendix C
SUBSTATION BUS CONFIGURATIONS

Most substation are organized similarly to one of four standard configurations. These are referred to as *bus configurations*. The aluminum cable or pipe typically used to make power connections among individual substation equipment is referred to as the *conductor.* The collection of equipment used to connect one or two transmission lines (depending on the bus configuration) into a substation is called a *position*. The main conductors that serve to connect positions together are called a *bus*. It is the configuration of the bus that gives a configuration its name. While there are many minor variations, the four bus configurations are called breaker-and-a-half, double-bus–double-breaker, double-bus–single-breaker, and ring bus.

An operating diagram is typically used to describe a substation. A single-line operating diagram gives a simplified schematic circuit diagram of the substation. It gives an overall view of what equipment is at the substation and how it is connected. It does not show the physical layout of the substation, although there is often a close correspondence and it is simplified in that the three phases are shown as a single line. Figure C-1 shows a simplified, operating, single-line diagram of a substation with a 500-kV switchyard in a ring bus configuration and a 230-kV switchyard in a breaker-and-a-half bus configuration. Note that the bus in each of these configurations serves to connect the transmission lines together from a node in the system. Also note that, in both configurations, there is a disconnect switch on each side of each circuit breaker. In this way, the circuit breaker can be isolated for performing maintenance. In the ring bus, any circuit breaker can be damaged and taken out of service without affecting the operation of the system. If two adjacent circuit breakers are taken out of service, the line between them will be isolated from the substation. If nonadjacent circuit breakers of the bus are taken out of service, the original node formed by the ring bus will be divided into two nodes and change the configuration of the network.

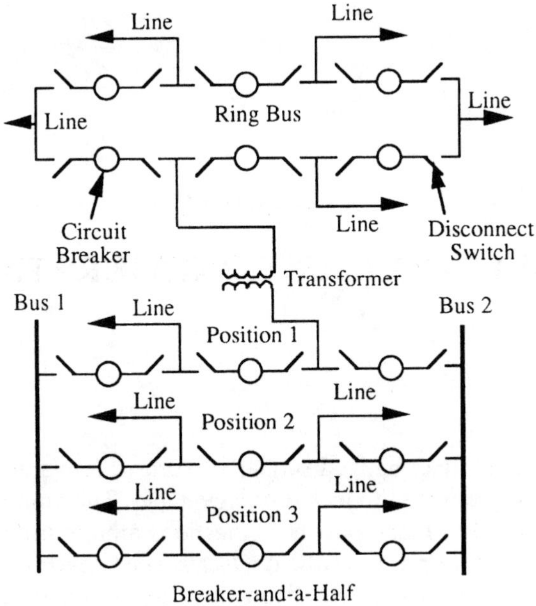

FIGURE C-1. Single-Line Operating Diagram with a Ring Bus and a Breaker-and-a-Half Bus.

Note: For simplicity, other equipment usually shown on single-line diagrams has been eliminated.

In the breaker-and-a-half bus configuration, each position services two lines. If any single circuit breaker in a position is taken out of service, there will be no effect on operations. If two circuit breakers are taken out of service, the line between them will be isolated from the substation. Note that each position has two lines and three circuit breakers, or a breaker-and-a-half per line, the source of the name for the bus configuration. This is generally the most common type of bus configuration for transmission lines. These diagrams are simplified in that other equipment, such as wave traps, potential transformers, and current-voltage transformers, have not been shown. In real operating diagrams, each element is assigned a number, usually associated with its position. While the numbering systems vary among utilities, for a given utility the numbering system simplifies identifying the location of the equipment in the switchyard.

Figure C-2 shows a simplified single-line operating diagram of a 230-kV switchyard in a double-bus–double-breaker configuration and a 60-kV switchyard in a double-bus–single-breaker configuration. In the double-bus–double-breaker switchyard, a circuit breaker in each position can be taken out of service without effecting operations. Note that this is the most expensive configuration, as each line requires two circuit breakers. However, it is not

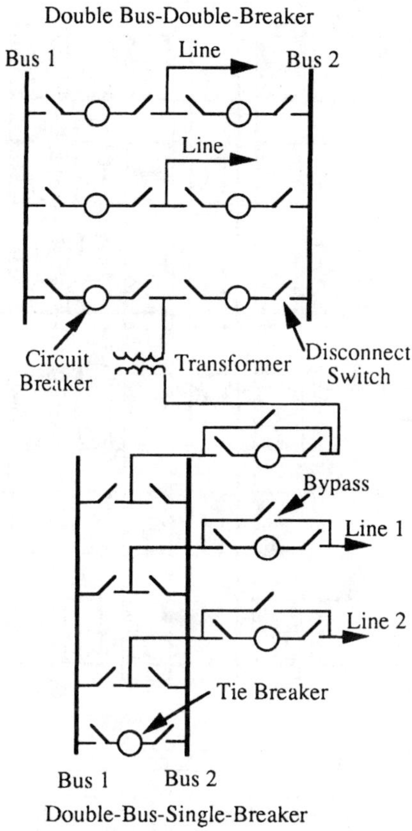

FIGURE C-2. Single-Line Operating Diagram with a Double-Bus–Double-Breaker Bus and a Double-Bus–Single-Breaker Bus.

Note: For simplicity, other equipment usually shown on single-line diagrams has been eliminated.

significantly more reliable than a breaker-and-a-half configuration, because the failure of circuit breakers is relatively rare (except in earthquakes).

In the double-bus–single-breaker configuration, there is one breaker for each line. To provide some measure of reliability, a tie-breaker is added across the busses. In normal operation, all lines are connected to Bus 1. If, for example, the circuit breaker on line 2 is taken out of service, the line is connected to Bus 2 and the tie breaker is used to connect Bus 1 to Bus 2. In this way, the tie breaker provides protection to Line 2. In this bus configuration, only one breaker can be taken out of service for all positions. In the other double bus configurations, one breaker in each position can be taken out of service. Note that in the double-bus–single-breaker configuration shown, each circuit breaker has isolation and bypass disconnect switches.

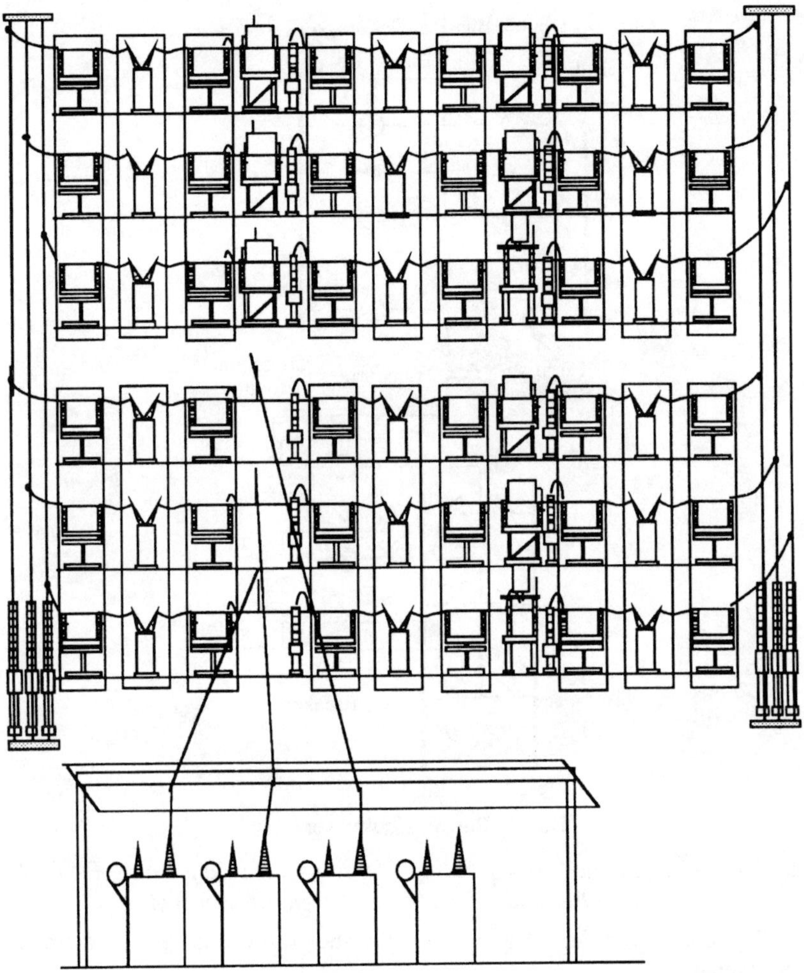

FIGURE C-3. Schematic Diagram of a Breaker-and-a-Half Bus Configuration with a Transformer Bank.

Figure C-3 shows a schematic diagram for two positions of a breaker-and-a-half bus. This diagram contains other substation equipment, such as wave traps and instrumentation transformers, not shown on the above diagrams. The schematic shows all three phases and pictorial representations of the equipment. While this diagram provides much more information about installation details and equipment configurations, for an understanding of the network configuration the simplicity of a single-line diagram is more convenient.

Appendix D
FURTHER RESOURCES

Algermissen and Perkins. 1976. U.S. Geological Survey Open-File Report 76-416.

American Concrete Institute. 1988. *Specification for the Design and Construction of Reinforced Concrete Chimneys.* ACI 307-88.

American Petroleum Institute. 1995. *Welded Steel Tanks for Oil Storage.* ANSI/API Standard 650-1993.

American Society of Chemical Engineers, Technical Council on Lifeline Earthquake Engineering. 1980. *Annotated Bibliography on Lifeline Earthquake Engineering.*

American Water Works Association. 1997. *Standard for Welded Steel Tanks for Water Storage.* ANSI/AWWA D100-1996.

Bolt, B.A. 1978. *Earthquakes: A Primer.* San Francisco: W.H. Freeman and Co.

Federal Emergency Management Agency. 1998. *1997 Edition: NEHRP Recommended Provisions for Seismic Regulations for New Buildings, Part 1: Provisions.*

Frankel, A., C. Mueller, T. Barnhard, D. Perkins, E. Lenendecker, N. Dickman, S. Hanson, and M. Hopper. 1996. *National Seismic-Hazard Maps: Documentation.* U.S. Geological Survey Open-File Report 96-532, June.

Hamilton, R.M. 1981. "Geologic Origin of Eastern U.S. Seismicity." In *Earthquake and Earthquake Engineering in the Eastern United States,* Beavers (ed.). Ann Arbor, Michigan: Ann Arbor Science.

McBean, R.P., L.E. Escalante, and A.W. Shively. 1983. "Intermountain Power Project Seismic Criteria." *Journal of Technical Topics in Civil Engineering* 109(1, April): 73–87.

Singh, S., M.K. Ravindra, and T.G. Longlais. 1981. "Seismic Design Criteria for Fossil Plant Structures." *Journal of the Energy Division, ASCE* 107(EY2, December): 261–275.

Appendix E

TECHNICAL COUNCIL ON LIFELINE EARTHQUAKE ENGINEERING PUBLICATIONS AND MONOGRAPHS

PUBLICATIONS

The Current State of Knowledge of Lifeline Earthquake Engineering. C. Martin Duke, ed., 1977.

Annotated Bibliography on Lifeline Earthquake Engineering, Munson Dowd, ed., 1980.

Lifeline Earthquake Engineering: The Current State of Knowledge, D. J. Smith, Jr., ed., 1981.

Advisory Notes on Lifeline Earthquake Engineering, William Hall, ed., 1983.

Guidelines for the Seismic Design of Oil and Gas Pipeline Systems, TCLEE Committee on Gas and Liquid Fuel Lifelines, Douglas Nyman, ed., 1984.

Lifeline Earthquake Engineering: Performance, Design and Construction, James Cooper, ed., 1984.

Lifeline Seismic Risk Analysis: Case Studies, Ronald Eguchi and C. B. Crouse, eds., 1986.

The Mexico Earthquakes: 1985 Factors Involved and Lessons Learned, Michael Cassaro and E. Martinez-Romero, eds., 1986.

Seismic Evaluation of Lifeline Systems: Case Studies, Leon R. L. Wang and Robert Whitman, eds., 1986.

Seismic Design and Construction of Complex Civil Engineering Systems, Michael Cassaro and James Cooper, eds., 1988.

Hyogo-Ken Nanbu (Kobe) Earthquake of January 17, 1995: A Post-Earthquake Reconnaissance of Port Facilities, Committee on Ports and Harbors Lifelines, Stuart D. Werner and Stephen E. Dickenson, eds., 1996.

MONOGRAPHS

No. 1, *Recent Lifeline Seismic Risk Studies*, A. Kiremidjian, ed., 1990.

No. 2, *Seismic Loss Estimates for a Hypothetical Water System*, Craig Taylor, ed., 1991.

No. 3, *Guide to Post-Earthquake Investigation of Lifelines*, Anshel J. Schiff, ed., 1991.

No. 4, *Lifeline Earthquake Engineering,* Michael Cassaro, ed., 1991

No. 5, *Lifeline Earthquake Engineering in the Central and Eastern U.S.,* D. Ballantyne, ed., 1992.

No. 6, *Proceedings of the 4th National Conference on Lifeline Earthquake Engineering,* Michael O'Rourke, ed., August 1995.

No. 7, *Critical Issues and State of the Art on Lifeline Earthquake Engineering,* Anshel J. Schiff and Ian Buckle, eds., October 1995.

No. 8, *Northridge Earthquake: Lifeline Performance and Post-Earthquake Response,* Anshel J. Schiff, ed., August 1995.

No. 9, *Seismic Design Guide for Natural Gas Distributors,* Peter McDonough, ed., August 1995.

No. 10, *Methods of Achieving Improved Seismic Performance of Communications Systems,* Alex Tang and Anshel Schiff, eds., September 1996.

No. 11, *Guide to Post-Earthquake Investigation of Lifelines,* Anshel J. Schiff, ed., July 1997.

No. 12, *Seismic Guide to Ports,* Stuart D. Werner, ed., March 1998.

No. 13, *Overcoming Barriers: Lifeline Seismic Improvement Programs,* Craig Taylor, Elliott Mittler, and Le Val Lund, September 1998.

No. 14, *Hyogoken-Nanbu (Kobe) Earthquake of January 17, 1995; Lifeline Performance,* Anshel J. Schiff, ed. September 1998.

No. 15, *Guidelines for the Seismic Evaluation and Upgrade of Water Transmission Facilities,* John M. Eidinger and Ernesto A. Avila, eds., January 1999.

INDEX